Francis Warner

Three Lectures on the Anatomy of Movement

A Treatise on the Action on Nerve Centres and Modes of growth, Delivered at the

Royal College of Surgeons of England

Francis Warner

Three Lectures on the Anatomy of Movement
*A Treatise on the Action on Nerve Centres and Modes of growth, Delivered at the Royal
College of Surgeons of England*

ISBN/EAN: 9783337270049

Printed in Europe, USA, Canada, Australia, Japan

Cover: Foto ©berggeist007 / pixelio.de

More available books at **www.hansebooks.com**

THREE LECTURES

ON THE

ANATOMY OF MOVEMENT

*A TREATISE ON THE ACTION OF NERVE-
CENTRES AND MODES OF GROWTH*

DELIVERED AT THE

ROYAL COLLEGE OF SURGEONS OF ENGLAND

By FRANCIS WARNER

M.D., F.R.C.S., F.R.C.P.

HUNTERIAN PROFESSOR OF COMPARATIVE ANATOMY AND PHYSIOLOGY IN THE
ROYAL COLLEGE OF SURGEONS OF ENGLAND

PHYSICIAN TO THE LONDON HOSPITAL, AND LECTURER IN BOTANY AT THE
LONDON HOSPITAL MEDICAL COLLEGE

FORMERLY PHYSICIAN TO THE EAST LONDON HOSPITAL FOR CHILDREN

LONDON

KEGAN PAUL, TRENCH & CO., 1, PATERNOSTER SQUARE

1887

PREFACE.

——•◦•——

THE purpose of these lectures is to show that, in studying the motor action of the brain, we are studying an integral portion of the body which is subject to the same laws as the rest of the body, and that the forces which, acting upon the brain, stimulate motor action are those which stimulate growth in other parts and other tissues.

It will be shown that motor acts may be described by referring to the parts moving, and the attributes of the movement, its time and quantity; while growth is described by referring to the parts growing, the time, quantity and kind of growth.

It will be demonstrated that in many cases the complex result following both motor actions, and acts of growth are due to the time and quantity of the component individual acts, and that the attributes may be controlled by physical forces.

Such considerations as these, when followed out further, may extend our knowledge as to the processes of evolution, and as regards motor actions they may give some explanation of the origin of many modes of expression of the action of mind.

The view is put forward that every vital act in a part of the body necessitates, not only a supply of pabulum, but also stimulation by some force acting upon it, each factor being capable of controlling its attributes and determining the result.

It seems to me desirable, as far as possible, to employ the method of describing the attributes of vital acts, in the belief that when we have more knowledge of the processes of nature we may thus be enabled to bring many phenomena within the scope of exact calculation. This mode of analyzing and describing movements enables us to give purely physical descriptions in place of using metaphysical terms, and also shows how they may be logically compared with other facts enabling us to make widespread analogies.

Many physical signs of physiological and morbid states will be given.

F. W.

24, HARLEY STREET, W.

*_** I am indebted to Mr. John Murray for permission to copy three illustrations from Mr. Darwin's works.

CONTENTS.

——◦◇◦——

LECTURE I.

Methods of biological study necessitate observation and thought.—Movements are signs of action.—Clinical signs of brain-states are mostly motor.—Actions, speech, postures are produced by movement.—Observation of movements more easy than their description and classification.

Movements described verbally, and by graphic method. — Tracings indicate time of action.—Postures also indicate brain-states.—Nerve-muscular apparatus, and nerve-centres. Antecedents and sequents of movements.

Time and Quantity the only intrinsic attributes of a movement.—A Posture is due to a balance of action in the nerve-muscular apparatus.—The part moved ; the time, quantity, antecedent, and sequence of the movement.—Parts of the body moving separately.

Separate action of nerve-centres, movements indicate their action : its time and quantity.—Time as an attribute of movement.—Combinations of movements ; their sequence ; their series.—The limit of possible combinations : formulæ.

Description of action as series of movements, or as series of combinations.—Examples : the relative advantage of each method.—The number of parts moving deduced from the number of combinations observed : formula.—Study of the expression of fright.

Metaphysical terms, and physical signs.—An action is composed of movements : a child seizing an orange.—Weeping a series of movements.—Movements in leaf of *Drosera rotundifolia*, in *Desmodium gyrans*, and in *Mimosa pudica*.— Importance of time of action.—Quantity an intrinsic attribute.—Ratio of action in nerve-centres indicated by Postures.—The straight extended hand : the Nervous-hand posture.—Hand in posture of rest : the Energetic hand.— Ratios of action in nerve-centres are determined by sights and sounds, etc.—Asymmetry follows lessened nutrition.— Quantity of movement determined by blood-supply, and stimulation by physical forces. — Movement lessened in Anæmia, starvation, and darkness.—Quantity of movement in plants dependent on light and heat.

Every movement depends upon the time or quantity of action in nerve-centres, which must be stimulated by some force.

Antecedents of movements—light, heat, sound, mechanical pressure, etc.—Sequents of movements ; confusion between antecedents and sequents . Bee visiting a flower.—Sir John Lubbock's experiments.—A child running.—Useful movements.—Impressionability indicated by movement.—Scientific descriptions.—Analogy between movements ; similar series of movements analogous in the parts, and in the attributes of action.—Imitation.—Repetition, or reversion of movements. —Movements ; automatic, instinctive, intelligent.—Respiratory movements ; co-ordinated movements ; reflex movements.—Classification.

The use of studying movements as signs of brain-action.— Metaphysical terms to be avoided.

LECTURE II.

The principles derived from the study of movements applied to conditions of growth.—A physical sign of function is a transmutation of force.—The same forces regulate movement and growth.—Growth and movement compared.— Intellect indicated by movement in an animal, but not in a plant.—Growth no sign of intellect.—Results of growth permanent, not so those of movement.—Kinetic and trophic function.—The part growing, its subdivision.—Growth a more complex study than movement.

Time of growth.—A posture is a condition of temporary equilibrium.—Proterogynous flowers: *Scrophularia;* protandrous flower of *Clerodendron.*—Time of various acts of growth.

Quantity as an intrinsic attribute of growth.—Description in terms of ratios of growth seen in a specimen, this is analogous to the study of a posture.—The forces which control postures are those which control growth.—Development of buds.—Contrast of a wild and a cultivated carrot.— Fruits.—Leaves.—Normal porportional growth : in a tissue ; in a young and in an older leaf.—Proportional growth of bilateral parts ; if unequal it leads to curvatures : curvature of ribs of foetus.—Curvatures causing nutation in plants.— Hyponasty and Epinasty.—Symmetrical growth.—Foetal growth.

Antecedents of Growth.—Action of light on growth of leaves and flowers, and in checking growth of internodes.—Mechanical pressure controlling growth, in bone, in a foot, in a palsied limb, in ulcers.—Heat in its climatic effects on growth.— Gravity causing curvatures in plants.—Sequences of proportional growth.—Effect of uniform extension of a surface upon holes made in it; foramina in pelvis ; fontanelle of infant enlarges with growth, then closes.—Hypertrophy ; its effects.—Effects of enlargement of guard-cells of a stoma. —Proportional growth in flowers, leading to cross-fertilization

LECTURE III.

Reversion.—Methods of observation deduced from the study of Chorea.—Description of the state of Coma.—Comparative study of Hemiplegia and Hemispasm as to parts concerned, the quantity and time of action.—Exhaustion of brain-power is physiological if bilateral, but it is called hysteria if one-sided.—Study of movements and postures in patients ; symmetry of action.—Study of the human face ; symmetry of action.—Description of a *nervous child* in terms of movements and proportional growth ; movements, postures, tooth-grinding, emaciation, etc.

Abnormal growth, often due to ratios of growth.—Hypertrophy from supply of pabulum.—Atrophy from pressure.— Hypertrophy from intermittent pressure ; from pressure on leaf-stalk of *Solanum Jasminoides.*—Nerve-force if deficient followed by atrophy in Anterior Poliomyelitis, in Hemiplegia; in optic tract when eyeball is lost ; in parts supplied by fifth cranial nerve.

Atrophy from disease is really due to defective stimulation. —Loss of eyes in Crustaceans from darkness, not from disuse. —Spontaneous movement not followed by hypertrophy ; voluntary movement followed by hypertrophy because it implies much stimulation from without.

Pathological conditions due to ratios of growth.—Signs of old age, diminished and increased quantities of growth : in organs, hair, prostate, and development of cancer.—Closure of openings in infant's skull ; in abdominal walls ; in auricular septum. ——Pathological conditions due to ratio of growth of tissues ; in strumous diathesis ; in morbid growths not malignant ; in plants.—Rickets ; bones curved owing to unequal bilateral growth, and in part to strains ; foramina distorted owing to absence of uniform extension in all parts. —Defects of proportional growth produce teratological specimens : a deformed chick ; hypertrophy of fingers ; multiple defects in a family.—Contrast of such cases with a kidney overgrown from surplus supply of blood.—A deformed fœtal skeleton.—Deformity of ears and other parts.

The seat of pathological changes as a basis of classification. —Analogy between isolated brain-centres and maculated skin-

LECTURE I.

THE STUDY OF MOVEMENTS.

B

A TREATISE

ON THE ACTION OF NERVE-CENTRES AND MODES OF GROWTH.

LECTURE I.

THE STUDY OF MOVEMENTS.

THE student of biology must ever be ready to observe new facts, and be prepared to collate them with other known facts, at the same time employing processes of thought and argument, in trying to see their relations and their meaning. The object of every scientific observation is to gain knowledge, and in conducting a scientific inquiry we must have some method of procedure.

It is proposed here to study movements as signs of physiological and pathological states, to seek explanations of the facts described, and to deduce from such studies methods which may serve as guides for other inquiries.

Analysis of the methods commonly employed in

clinical observations of the nerve-system shows
that we are accustomed to observe movements, and
results of movements, as physical signs. Movement is one of the most constant signs of life ;
looking at a man we observe his body, its form
and colour, its parts and their proportions, but
soon our attention is attracted by his movements,
his actions, and by his speech, and his postures,
which are results of his movements.

It was particularly when conducting an inquiry
as to the signs of development in children, and
their potentiality for future mental and moral
culture, that I was led to follow the lines of
investigation which I now have the honour to
put before you. It was easy to see movements
in man, it was more difficult to determine how
they might best be described, and studied in a
systematic manner. Having taken verbal descriptions of various movements, it appeared convenient
to employ the graphic method of record, a method
which has yielded so much information in the
investigation of physiological problems. Tracings
of various movements were obtained; the details
of the methods employed, and the apparatus used,
have been published in the *Journal of Physiology,*
vol. iv., No. 2 ; vol. vii., No. 4.

Fig. 1.—Tracings of the spontaneous movements of an infant's hand during fifteen minutes.

Tracings, or permanent records of movements, having been obtained in place of verbal descriptions, it became necessary to analyze and classify them, and to determine their significance. Tracings of movements indicate their time and frequency.

While engaged in the attempt to describe the movements seen in patients, it became evident that the postures often correspond with certain physiological or pathological conditions. Postures are the results of movement; they are due to states of equilibrium of muscular action.

We speak of observing movements in man. A movement involves a part moved, the muscles moving it, and a nerve-apparatus stimulating the muscles to contract; we observe the subject, the man, noting his movements. The nerve-muscular apparatus produces the movement, which is thus a sign of the action of the nerve-centre corresponding. Having seen the movement, we may proceed to try and determine its antecedents and sequents. In observing a movement, or the action of a nerve-muscular apparatus, we may note its time and its quantity; these are its only intrinsic attributes or characters. In observing postures of the body, we see the effects of the balance of pieces of the nerve-muscular apparatus in the condition of temporary equilibrium; that is, we observe relations as to quantities of action.

It is then suggested that we may observe movements in man, as signs of the action of the nerve-muscular apparatus, and that concerning each such movement we must consider—

The part moved.

The time of movement.

The quantity of movement.

The antecedents of the movement.

The sequents of the movement.

We shall proceed to inquire as to what points are observable under each of these heads.

As to *the part moved*, the part of the body seen in motion. Many parts of the body can move separately; they have been named by the anatomist, and their movements have been described in anatomical language; one or many parts of the body may be observed in motion at one time. We may readily suggest five classes, or groups of parts of the human body as worthy of study : the members; small parts and large parts; proximal and distant parts; bilateral parts of the body; parts moved by each group of muscles supplied by a special cranial nerve.

Probably any group of muscles able to act alone, has its own nerve-centre as a unit or locus of nerve-tissue. Thus, the individual digits can move separately, and the centres producing these movements are probably capable of separate action.

The correct identification of the parts observed is of great importance to our further arguments, the movement cannot be defined without reference to the parts seen in motion. The movement of a part is to us a sign of the time, and the quantity of action of that portion of nerve-tissue which supplies the force producing the movement of the part, or, as we commonly call it, the corresponding nerve-centre. Synchronous movements of several parts of the body, each such part being capable of separate movement, implies the synchronous action of several loci of nerve-tissue.

As previously stated, the attributes of a nerve-muscular movement are its time and its quantity, and these are its only intrinsic characters.

The time of the observed movement of a part is an attribute or character of extreme importance, for the time of the movement greatly affects its sequence, as will be shown. The movements of a single part of the body may be represented by a curved line, which indicates its time, rate, and frequency, and also that of the action of the nerve-centre corresponding; a line may thus represent a series of movements of a part. The movement of a single part is a comparatively simple fact. When several parts are observed moving, their actions may be synchronous or not so; if they be all synchronous the same act is repeated over again on each occasion, as in walking; if the movements be not synchronous, they may occur in various combinations, each combination of movements having its own sequence or producing its own effect. It is a simple matter to see the index-finger flex over and over again; the time of the various movements producing writing is a complex fact; here the result of the movements depends upon the time and the quantity of the individual movements.

The special combinations of movements that occur are simply the necessary result of the time of the action of each of the moving parts. It may prove a matter of great interest to study the description and the outcome of some combinations of movements.

We must now consider a few points on some abstract questions, and then we shall see how we are enabled to give descriptions of facts, and in some cases to give explanations of them. When two or more parts are seen moving, there is a series of movements of each part, and a series of combinations of movements. If we consider the

five digits, simply with regard to their being in a
state of absolute rest, or in a state of motion of
any kind, then thirty-one possible combinations of
movements may occur. The digits may be called
A, B, C, D, E; the possible combinations are A, B,
C, D, E, AB, AC, AD, AE, BC, BD, BE, CD, CE,
DE, ABC, ABD, ABE, ACD, ACE, ADE, BCD,
BCE, BDE, CDE, ABCD, ACDE, ABCE, ABDE,
BCDE, ABCDE. If n represent the number of
moving parts observed, it is possible to have $2^n - 1$
combinations, or different coincidences of movements.

In observing movements of the digits two modes
of description may be employed; the series of
movements of each digit may be described, or the
series of coincidences or combinations of movements seen may be recorded. Either plan may be
used, as convenience may dictate. In studying
the movements of the digits, we may consider
flexion only. Two modes of description may be
used; the movement of each digit may be described, or the successive combinations of movements of the several digits may be described.
Thus, it may be said that each of the five digits
flexed at the same moment, or it may be said that
the hand assumed the convulsive posture.

In describing what is seen according to the first method, we speak of five parts, indicating how each moves. The hand is observed with each digit extended, then every digit flexes; this is

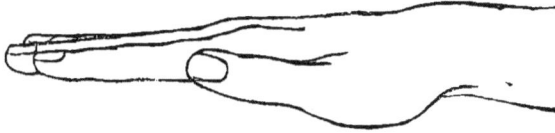

Fig. 2.—The Straight Hand.

followed by extension of each digit, next the thumb alone flexes. According to the second mode of description we may say that the hand is seen

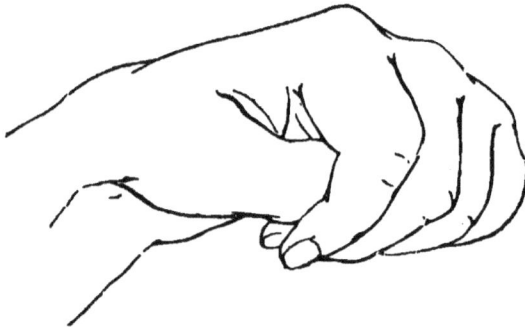

Fig. 3.—The Convulsive Hand.

in "the straight extended posture;" this is followed by "the convulsive hand;" next "the straight extended hand" is again seen; followed by "the straight extended hand with the thumb drooped."

The first method can always be employed when we are able to define the parts concerned in an action. The second method is often less laborious, but its employment may lead to errors in making the attempt to analyse the meaning of actions. To describe all the movements of the digits as suggested according to the second method, we must be prepared to describe thirty-one combinations.

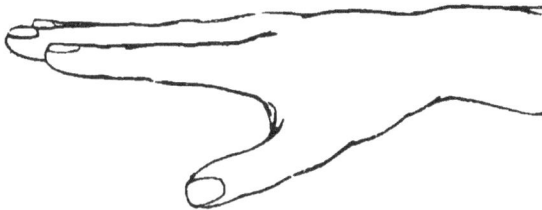

Fig. 4.—The Straight Extended Hand with the Thumb drooped.

Each special combination of movements is due to the time of the movement of each of the parts, and is to us a sign of the time of the action of the nerve-centres corresponding. The five digits flexing synchronously, or asynchronously, in all possible combinations, give us thirty-one forms of expression of the state of the nerve-centres.

There may be examples in which an observer looking at a number of movements is satisfied that he has seen all the possible combinations of movements, and has enumerated them, but does not know how many factors are concerned in producing those combinations. In such a case let A equal the number of combinations enumerated; X equal the number of factors concerned in producing those combinations of action; then $A = 2^x - 1$ and $X = \frac{\log. A + 1}{\log. 2}$.

It may be said of a man that he was frightened by the sight of a dog. This statement may be founded upon the observation of the expression of fright, and one among the points in this particular mode of expression is, the combination of movements of the digits, most easily expressed verbally by naming the posture assumed ; hence we consider the hand posture as a physical sign of the (mental) brain state. Speaking in other words of what we see, we may describe the movement of each of the parts as equal and synchronous extension of all the digits, and of the wrist, the extensor centres showing more force than the flexors. It may be noted that in the case supposed, the time and the quantity of the action of the nerve-centres concerned may be determined by the sight of a dog. To speak of fright apart from its physical expression is to use a metaphysical phrase. It may be convenient, and save trouble, to use such metaphysical phrases, but it is not a scientific method.

In observing the frightened man, we may note his movements and study them. Finding it laborious to describe the movements of each separate part, we use verbal expressions descriptive of combinations of movements; we say he presents "the hand in fright," etc.

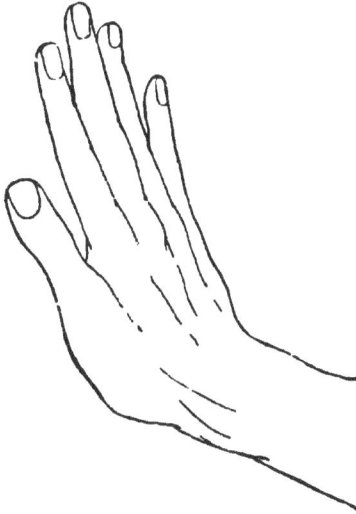

Fig. 5.—The Hand in Fright.

We may speak of a man performing some action, as writing, walking, carving, etc. What we commonly call an action, whether we speak of it as spontaneous, reflex, or voluntary, is an observed combination or series of acts or movements. It is important for our purposes to indicate the compound character of many actions, showing that they are composed of separate movements, an example or two will serve to illustrate this.

A child sees an orange; he seizes it, and eats it. This action seen in the child is a series of acts or movements, each of which can be performed separately, while the nature of the action depends upon the time, and the quantity of each of the component acts or movements. The humerus is raised and directed towards the orange; the elbow is extended and the fore-arm pronated, bringing the hand over the orange; the digits extend, and finally close on the object, thus seizing it, etc. It is the sight of the orange, the impression produced by light reflected from the orange, that is the antecedent of this series of movements, controlling their time or order, and the quantity of each movement. We are not here discussing how the mechanism of the child has come to be what it is. We shall study the antecedents and sequents of series of movements presently.

Sir Charles Bell * gives a description of *weeping*, mainly as a series of movements : " The lachrymal glands are the first to be infected; then the eye-

* "The Anatomy and Philosophy of Expression," third edition, p. 149.

lids ; and finally, the whole converging muscles of
the cheeks. The lips are drawn aside, not from
their circular fibres relaxing, as in laughter, but
from their being forcibly retracted by the superior
influence of their antagonistic muscles. Instead
of the joyous elevation of the cheeks, the muscle
which pulls down the angle of the mouth, *triangu-*

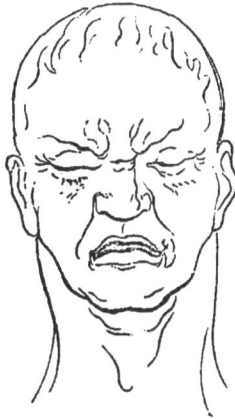

Fig. 6.—Weeping.

laris oris, is more under influence, and the angle
is depressed. The cheeks are thus drawn between
two adverse powers: the muscles which surround
the eyelids, and that which depresses the lower lip.

"The same cause which drew the diaphragm and
muscles of the chest into action in laughing is per-
ceived here. The diaphragm is spasmodically and

irregularly affected; the chest and throat are influenced; the breathing is cut by sobbing; the
respiration is hurried, and the expiration is slow,
with a melancholy note."

A series of movements having a definite ante-

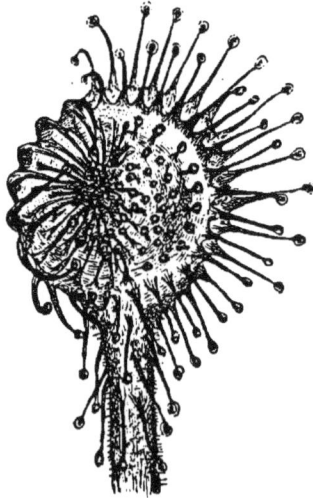

Fig. 7.—Drosera rotundifolia.
Leaf (enlarged) with the tentacles on one side inflected over a
bit of meat placed on the disc.

cedent and sequence, is seen in the case of the
tentacles of the leaf of *Drosera rotundifolia* when a
fly has become attached to one of them.

Mr. C. Darwin, in his work on " Insectivorous
Plants," p. 4, describes the leaves of *Drosera
rotundifolia* in detail.   The whole upper surface

c

of the leaf is covered with gland-bearing filaments, or tentacles. The glands are each surrounded by large drops of extremely viscid secretion, which, glittering in the sun, have given rise to the plant's poetical name of the sun-dew. The tentacles on the central part of the leaf are short and stand upright, and their pedicels are green. Towards the margin they become longer and more inclined outwards.

If a small organic object be placed on the glands in the centre of a leaf, these transmit motor impulse to the marginal tentacles. The nearer ones are first affected and slowly bend towards the centre, and then those further off, until, at last all become closely inflected over the object. This takes place in from one hour to four or five or more hours.

When an insect alights on the central disc, it is instantly entangled by the viscid secretion, and the surrounding tentacles after a time begin to bend, and ultimately clasp it on all sides. If an insect adheres to only a few of the glands of the exterior tentacles, these soon become inflected and carry their prey to the tentacles next succeeding them inwards; these bend inwards, and so onwards, until the insect is ultimately carried by a sort of rolling movement to the centre of the leaf. Then, after an interval, the tentacles on all sides become inflected, and bathe their prey with their secretion, in the same manner as if the insect had first alighted on the central disc.

The *Desmodium gyrans* affords another example.

The leaf of *Desmodium gyrans* is unequally pinnate, having a large leaflet at its extremity,

Fig. 8.—Desmodium gyrans, after Balfour.

and two small lateral leaflets. During the day the large leaflet rises with a slight lateral motion; these movements are slow. The lateral leaflets, on the other hand, constantly exhibit a jerking motion, by which they first approach each other, and then retire, the length of time required to complete their movements being about three minutes when the plant is vigorous and exposed to bright light.

The movements of the leaves of *Mimosa pudica* are well known.

Professor Balfour * gives a good description of the leaves of *Mimosa*. He says—

"The leaf of *Mimosa* is compound and bipinnate. The small pinnules or leaflets are expanded horizontally when the plant is in the light, and in its natural state ; but when it is in darkness, as well

Fig. 9.—Mimosa pudica.

as when its leaves are touched or irritated, the pinnules fold upwards, so as to bring their upper surfaces into contact, and at length the petiole is depressed, so that the entire leaf falls down. When light is introduced, or when the irritation is removed, the leaflets gradually unfold, and the leafstalk rises. If two of the leaflets at the extremity

* "Class-book of Botany," third edition, p. 495.

are touched, or are irritated by heat from a lens or by electricity, without agitating other parts, they fold upwards, and a similar movement takes place in the adjoining leaflets in regular succession from the apex to the base of the petiole. The irritation is also communicated to the neighbouring partial petiole, the leaflets of which fold in a reverse order, namely, from base to apex. The movements may be propagated until the partial petioles converge and fall down; and finally the general leaf-stalk is depressed. If the lower leaflets are first irritated, the foldings take place from the base to the apex of the petiole; if the middle leaflets are touched then the foldings occur on each side."

The following experiments with the *Mimosa* have a bearing on the causes of impressionability; they were performed by Mr. M. Voss, who kindly communicated them to me.*

"Some seed of the sensitive plant (*Mimosa pudica*) was set to grow, and at a moist heat of about 90° Fahr. it soon germinated. Before the compound foliage growth had commenced, the seedlings were potted off into different earths and sand.

* See *British Medical Journal*, February 25, 1882, "Analogy between the Movements of Plants and the Muscular Movements of Children, called Chorea."

Those planted in a soil of two parts of decayed vegetable mould to one of sand grew more vigorously both in height and foliage than the others; and, after two months' growth, they were much less sensitive than others planted in two-thirds of silver sand and only one-third of leaf-mould. One or two plants were grown entirely in silver sand. These showed extreme sensitiveness to the slightest touch; even a breath of air, or the slightest jerk of the pot in which they grew, caused all the foliage to shut up."

Some lessons have been taught as to the time of actions. Time is an important attribute of an action, but the force stimulating an action is not necessarily immediately antecedent. Bringing a light in front of the eye is immediately followed by contraction of the pupil; such act is therefore called a reflex action. Even here it has been demonstrated that there is a measurable interval between the stimulus and the contraction. Seeing an object is the cause of its being talked of next day, though the interval may be long. These examples of delayed expression of impressions are often considered signs of mental action.[*]

Mr. C. Darwin says of the tendrils of *Dicentra*

* See "Physical Expression," p. 250.

thalictrifolia, " The terminal branches when lightly rubbed with a twig became curved in the course of from thirty to forty-two minutes, and straightened themselves in between ten and twenty hours." * The friction upon the tendril produced an impression, the movement occurred some time later.

The second attribute of a movement that we have to consider is its *quantity*. Every movement in the human body is the motion of a material part, and is capable of estimation in terms of mechanical work done. The quantity of motor-force is in some degree a sign of the quantity of action exerted by the nerve-muscular apparatus. The strong clenching of the fist indicates a large quantity of force passing from the nerve-centres to the muscles, producing flexion. When the fingers are relaxed this signifies less energy discharged from the centres.

In two or more parts we may note the ratio of motor force, as indicative of the ratio of action in the nerve-centres corresponding. Look at the open straight extended hand (see fig. 2) this posture indicates temporary equilibrium in the nerve-muscular apparatus for flexion and extension; the posture may change, the fist being clenched (fig. 3). This change of hand posture is a sign of a change

* Climbing Plants," p. 125.

in the ratio of force acting in the apparatus for flexion and extension, the force of flexion being increased. Thus we see the outcome of the balance, or ratio of quantities of motor-force, demonstrated by the observation of postures.

I have elsewhere described * "the posture of the nervous hand" as a typical posture pathognomonic of the neurotic condition, often seen in weak or

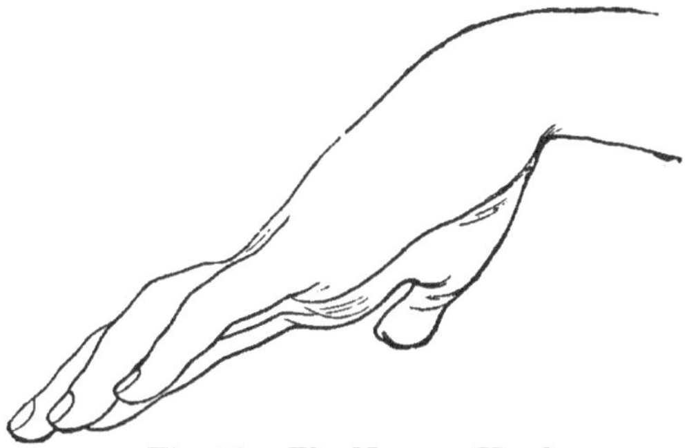

Fig. 10.—The Nervous Hand.

choreic children. The wrist is flexed, the metacarpo-phalangeal joints are over-extended, the thumb being likewise extended. In the wrist, flexion is stronger than extension, in the metacarpo-phalangeal joints extension is the stronger. The important element in this example is the quantity of motor or kinetic action in each nerve-muscular apparatus, the posture is determined by their ratios.

* See "Physical Expression," p. 162.

In a man, observe the hand when in the position of rest; all the joints are slightly flexed. Now let

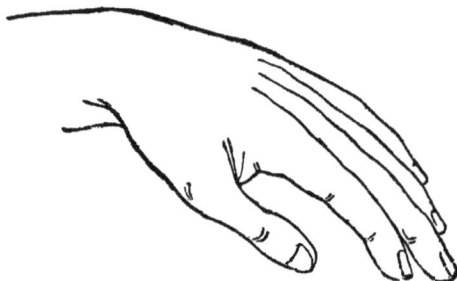

Fig. 11.—The Hand in Rest.

the sight of a horse surprise him, his hand will probably assume for a moment " the energetic posture," the wrist being extended, and the fingers remaining

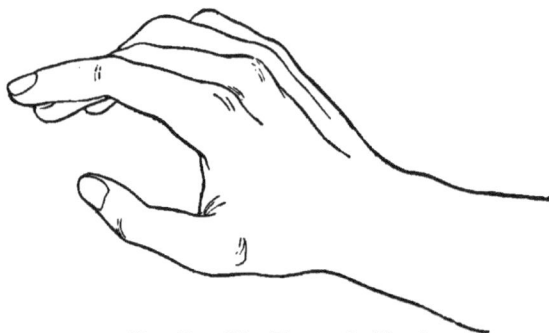

Fig. 12.—The Energetic Hand.

semi-flexed. This change of posture, this change in the ratios of quantities of action in the nerve-muscular apparatus, is sequential to the sight of the horse; that is, it is sequential to the incidence of light reflected from the body of the horse.

We may now summarize, and make the following statements. Postures are due to the ratios of the force exerted by opposing portions of nerve-muscular mechanism. Postures are, therefore, signs of the ratios of force or action. Postures are ratios of action in the parts considered, which can be changed by incident forces, such as light, mechanical irritation, etc.

We shall see in the second lecture that legitimate analogies may be made between postures and proportional growth, both being related to quantities of action ; the former with quantities of kinetic or motor-action, the latter with quantities of trophic action. In each case the ratios may be controlled by such forces as light and heat, etc.

The whole quantity of movement seen in man may depend upon the quantity of blood supply, and upon the amount of the stimulation of the subject. Among the stimuli which excite movements we may enumerate light, sound, heat, mechanical touch, etc. Both the supply of blood, and stimulation are needed to produce movements; each factor may be considered separately.

In anæmia the movements of the patient are slow

and feeble ; starvation lessens the motor-power and the quantity of movement. In a starved or anæmic child, or in an infant with a patent fontanelle, shrunken from want of blood-supply to the brain, movements are much diminished in quantity ; when the infant is restored to health, the normal amount of movement returns. This shows that the nerve-muscular mechanism requires a supply of good blood, in order that it may produce a sufficient quantity of movement. Diminished stimulation by external forces, even if the circulation be good, results in lessened movement. In darkness but little movement occurs in man ; the sight of an object often causes the hand and head to move towards it, and this may be followed by other series of movements. The sight of a toy may increase the quantity of movement in a child, the head and eyes turn towards the object and as soon as it is well within the field of vision, the hand is stretched out and seizes it.

The quantity of movements in plants is mainly determined by light and heat.

It is convenient in this place to enunciate the proposition that every movement depends on some change in the time or the quantity of action in the nerve-muscular apparatus producing it, and that every such change in the apparatus must be preceded by some stimulus acting on it immediately antecedent, or more remotely antecedent. Various examples have been given ; more will follow.

We have considered the two essential intrinsic attributes or characters of movement, viz. its time and its quantity. We must now pass on to consider its antecedents and sequents.

Important *antecedents of movements* are the forces which, acting upon a man, stimulate and control his movements. Among examples of such forces we may mention the following :—light, heat, sound, mechanical irritation or touch, etc.

Light is a common antecedent of movements. The sight of an object may cause weeping, a series of movements that has been described. Weeping may be caused also by cutaneous and auditory impressions. Light is a necessary antecedent to many of the movements of animals and plants.

Mechanical touch may be followed by a movement or a series of movements ; touching a man may be followed by rotation of his head, or by movements of his head, eyes, and upper extremity. Ants appear to intercommunicate by touching one another's antennæ. In the leaf of the *Drosera*, if a tentacle is touched by a fly, this is followed by a bending of all the tentacles towards the point first touched.

The *antecedents* or causes of movements are the stimuli which produce the movements; these we have spoken of already. Such antecedents often being physical forces are more easily understood, and seem more familiar than the complicated sequents which may follow.

We pass on to study the *sequences* of movements. Every movement must have some sequence or effect. It may produce an impression on the subject producing the movement, or it may be expended in physical force. The result of a movement in an animal may be to supply it with food or simply to move the air or objects around, expending itself in physical energy. The sequences of an action or movement must be carefully separated from the cause or antecedent. It is not uncommon to see examples of confusion between the antecedents and sequents of an action. It is said that the bee visits the flower to get the honey; that is to imply that the honey causes the visit. What is seen is, that the bee gets the honey after visiting the flower. It is an assumption to say that the bee knows that he will get honey; it may be true that he obtained honey on former visits to flowers. If this assumption be laid aside, and the results of experiments be looked to, it will appear that the colour of objects—

that is, rays of light—from the flowers are necessary antecedents to the visits of the bees; this is a potent cause of the visits. Such mistakes are apt to occur when we depart in the least degree from the method of simple direct observation with our senses.

Sir John Lubbock,* in his experiments on bees, demonstrated that their visits in search of honey are largely guided by the colour of the object where the honey is found. Among other experiments is the following. Slips of glass, all similar in size and form, had coloured paper pasted on them respectively—blue, green, orange, red, and white. These were placed upon the lawn in a row, about a foot apart, and on each was placed a second slip of glass with a drop of honey. Plain slips of glass, each with a similar drop of honey, were also placed with them. The greatest number of visits were paid to the blue glass, the smallest number to the plain glasses. Clearly, then, the colour had much to do with determining the visits of the bees.

As further illustration of the importance of discriminating between the antecedent and sequence of actions, take the following examples. It may be

* "Ants, Bees, and Wasps," International Science Series, p. 304.

said that the child puts out his hand and seizes the orange that he may eat it. Here the case is presented as if eating were some kind of cause or antecedent, to seizing the orange. Confusion on such points is, I think, common and very obstructive to the advance of scientific knowledge. The child runs to meet his father after hearing his voice; sound is the immediate antecedent of the child's movements, meeting his father is an observed sequence.

It will be granted that the forces which control the time and quantity of the acts composing a series of movements, are really the antecedents of the results of that series of movements.

The sequences, outcome, or results of movements are often said to be useful, intelligent, voluntary, purposive, etc. What is meant by the outcome of a movement being useful ? What is the connotation of the term useful expressed in terms of physical criteria that can be observed ?

Concerning the meaning of the word useful as applied to the outcome of an action or movement, the primary idea seems to be, that the outcome of the action produces some physical change that can be observed. A movement which brings food to the subject is said to be useful, partly because it produces some effect or impression on the subject. The movements of deglutition, respiration, and all the movements of organic life are useful, because, among other things, they impress the subject considered. We do not say that all acts which produce an impression are useful, but to be useful there must be survival in the subject producing the act, or in others considered at the same time. We insist that the usefulness is not an intrinsic character of any movement, but a relation of its outcome to surrounding objects.

The following are useful actions, known to be so by their physical effects :—seizure and conveyance of food; directing the eye towards an object; movements tending to reproduce the species. In plants we may cite, movements towards the light; movements of stamens ; movements due to geotropism, etc.

Impressionability as a property in man may be indicated by movements. If the movements of a man are regulated and controlled by light, he is impressionable to the action of light. This is known to us when a certain action follows the sight of an object, examples of which have been given; in coma the man is not thus impressionable. Similar statements might be made with regard to impressionability to sound, or touch, etc.

I have shown that in deep anæsthesia from chloroform,* or in coma from alcoholism, or in the profound sleep of infants, the loss of associated movements of the eyes may be complete. If in an adult deeply under chloroform, the eyelids be gently raised, the pupils will be seen minutely contracted, often to a pin-point, the eyes having at the same time lost the parallelism of their axes. One eye may move upwards or outwards while the other remains quiet, or moves in a different direction, or at a different pace, thus causing a temporary and varying strabismus. Usually these movements are confined to the horizontal plane, less commonly the eyes assume a different level, one being in the horizontal plane, while the other is turned downwards. It is noteworthy that the average continuance of the eyes in the horizontal plane of the axes of the orbits, is in accordance with other examples of involuntary movements of the eyes, *e.g.* nystagmus, and irregular jerking movements of the eyeballs. These movements I have frequently observed in the healthy subject, and seen that though they occurred thus irregularly while in coma, the action of the pupils and the co-ordination of the eye-movements were completely

* See paper, *British Medical Journal*, March 10, 1877.

D

restored on recovering consciousness. In the profound sleep of an infant in its mother's arms the
same loss of association of movements occurs, but
at the moment of waking the pupils dilate, and co-
ordination of movements is restored; the child
must be profoundly asleep to allow of the eyelids
being raised without awaking it.

CASE I.—At the Children's Hospital, Birming-
ham, in 1876, I saw a girl three years of age the
subject of permanent hemiplegic paralysis, following convulsions in infancy. She had never spoken,
was constantly dribbling, and idiotic in manner.
There was an occasional loss of the parallelism of
the eyes; one would remain at rest, while the other
wandered inwards or outwards ; this was a chronic
condition in the child, who was suffering from no
acute disease. The symptom appeared due to the
defective condition of the brain.

CASE II.—Through the kindness of my friend
Dr. Fletcher Beach, I had the opportunity in 1877 of
seeing in the Clapton Asylum a microcephalic idiot
of very low development, in whom the eyes almost
habitually wandered about independently of one
another.

In very weakly infants, not the known subjects
of brain-disease, the loss of associated movements

may be very distinctly seen at times while awake and sucking at a bottle. In cases of meningitis, and other conditions of coarse brain-disease, the same condition is sometimes seen.

The proposition will be granted that all scientific descriptions should be given in terms capable of observation by physical means. In attempting to carry out this rule it is often convenient to describe the attributes of time and quantity of the action, then to note its antecedents and sequents, at the same time clearly identifying the subject in which the act is seen. In giving descriptions of actions, especially as to their cause and effect, great care must be taken to be always logically correct in discriminating the antecedents from the sequents. In making comparisons or analogies between movements, scrupulous logical precautions must be taken to ensure that only like things, or like attributes are compared, quantity with quantity, and time with time.

We will now examine the grounds upon which we may say that certain movements, or combinations or series of movements are similar. If on two or more occasions, there be movements of the same parts of the body, the action of each part respectively being alike in time and in quantity on each occasion, then the two series of movements are similar.

If in two men the movements in like parts, strictly correspond as to their time and quantity of action, then the two series of movements are similar. Thus actions, or series of movements in the same or in corresponding parts, are identified as similar on account of their intrinsic attributes, time and quantity.

In a body of soldiers at drill, or in a school of children, imitating the manual exercises of the teacher, the movements are similar in each individual. An awkward recruit when marching with his company may keep time, but lift his feet higher than the others; his movements are similar to theirs in time, but not in quantity.

Similar series of movements may recur in a man, or be repeated, each such occurrence after the first is a repetition or reversion of the action or series of movements. A trick or habit in a man may revert in his children. We here define similarity of actions as an observed fact, not as a question of relation to antecedents or sequents of the action.

Movements may be classified in many different ways; they are sometimes spoken of as "automatic," "spontaneous," "voluntary," "purposive," "purposeless," "useful," "co-ordinated," "intelligent," "instinctive." In employing such terms we do not directly indicate facts capable of observation by the senses; they are not terms expressing characters dependent upon the intrinsic attributes of the movements observed.

Automatic or spontaneous movements.—Such terms imply that no special antecedents have been determined for such movements controlling their attributes; they are supposed to have no special relations to the surroundings. Athetoid movements are automatic; they recur, usually in similar series, whenever any impression is made upon the subject.

Instinctive movements.—Probably it is here necessary that the nerve-system should have a certain structure, due to its inheritance, such that when stimulated by certain surrounding forces the movements resulting should be regulated as to their attributes by those forces, and the sequences of such acts produce some impression upon the subject directly or indirectly.

Purposive and intelligent movements are probably always controlled by forces surrounding the subject, or by impressions previously made; the sequences of such movements rather than their antecedents possess the characters of intelligence. It seems that the purposive or intelligent character of a series of movements is not due to its intrinsic attributes, but to its sequents.

The respiratory movements are considered very automatic when occurring in regular and uniform order of time and quantity, but when the sight and sound of various objects modifies them, to form speech, laughter, etc., then they are signs of mental action. All expression of mind and the mental states is by series of movements, and these of course have their intrinsic attributes observable in the body.

Co-ordinated movements.—This term indicates a series of movements having a special character dependent upon its intrinsic attributes, the time and quantity of the component movements. It is also an essential character of co-ordinated movements that the attributes of the series are determinable by sights, sounds, and other forces acting upon the subject. A child seizing an orange presents a co-ordinated series of movements.

Movements may be classified according to *the parts moving.* Thus we speak of symmetrical movements; movements of the head, tongue, face, etc. ; also movements of large and small parts.

Reflex movements.—Here we refer to the action of a special piece of nerve-muscular apparatus, an afferent nerve-tract, and the stimulus acting upon a sensory surface followed by movement.

Movements may be classified *according to their results*, as respiratory, acts of deglutition, etc. It has been shown that useful movements derive their characters mainly from their sequences and surroundings.

For the purposes of scientific classification, it is very desirable that movements should always be described according to the parts moving, their intrinsic attributes, the antecedents and sequents or outcomes. All these characters are capable of direct observation.

It has been said, What is the good of these studies of movements ? What practical advantages are likely to result therefrom ? Nerve-muscular movements are physical signs of brain-action ; with but few exceptions they are the only signs of brain-action, and the only physical signs by which we know anything of the action of Mind.

Describing facts in man as we see them, in terms of movement, is probably the best mode of attempting to do away with such metaphysical terms as " voluntary," " intelligent," " purposive," in scientific descriptions. Facts must be described in physical terms before we are likely to find out their causation.

Mr. C. Darwin's investigations on the movements of the parts of plants, and his methods of recording them, led to a further and unexpected knowledge of the sensitiveness of plants, and the constant movement of all parts of the young seedling.

We are not here concerned with clinical practice, but I have elsewhere been able to suggest certain postures, and movements affording indications of disease. Many of these signs have proved of much value in determining the state of nerve-system in children as seen in schools, when questions cannot be asked, only signs observed.

These studies teach some important lessons, and suggest important propositions which cannot be followed out here.

The only intrinsic characters of movements are time and quantity ; such characters as intelligent, useful, voluntary, are relations to antecedents, sequents, or surroundings.

LECTURE II.

THE STUDY OF MODES OF GROWTH.

LECTURE II.

THE STUDY OF MODES OF GROWTH.

In making a systematic study of movements in man, we study one function in one class of subjects, that is, the motor or kinetic function in the nerve-muscular apparatus of man. We observe the movements as signs of the time, and quantity, of action in the loci of nerve-tissue which produce them; thus we learn much concerning vital action in one kind of tissue.

If the methods and principles found useful in studying this one function in one tissue, have been properly considered, they are likely to be useful in studying other functions in other living subjects.

In studying movements we learnt to note the parts of the subject capable of acting separately. We learnt next to study the intrinsic attributes of each movement, or kinetic act, its time and quantity of action, and also the sequences and antecedents of the movement. Further, we found it possible to enunciate the proposition that every movement, or nerve-muscular act, probably necessitates an antecedent force acting upon, or in, the subject, as well as a supply of blood. We shall presently enunciate an analogous proposition with regard to growth, nutrition, or any vital act. We take it for granted that any act, or physical sign of function, is a transmutation of force, not a creation of force.

Having studied the nerve-muscular apparatus of man as our special subject, and deduced certain generalizations and principles therefrom, we shall proceed to apply these to the study of various problems in comparative anatomy and physiology.

In this era, when successful efforts are being made to apply the doctrines and generalizations of the hypothesis of evolution to the study of facts at large as seen in living beings, it seems reasonable to seek for new methods of inquiry in harmony with such doctrines.

Interesting and important sequences of brain-action have been seen to follow the impress of forces acting upon the brain; we shall find that the same forces influence conditions of development and growth. We found that many complicated actions of the brain are relations of the time and quantity of action in the units of the brain; we shall find that many complicated phenomena of growth are determined by the time and quantity of trophic action in the units of living tissues.

In describing acts of growth we shall endeavour to give anatomical description of the parts growing, and note the time, the quantity, and the kind of growth or change occurring in each part, also its antecedents and sequents. In such studies we shall observe living things, studying trophic action in place of kinetic action. We shall compare or make analogies between different processes of nature, in the same subject or in different subjects, and the criteria of resemblance or difference will be the time, quantity, and kind of the action seen. Thus we want to make analogies between motor acts and acts of growth.

We may here stop for a moment to compare growth and movement, as examples of vital action in living beings; we want to show their general resemblances and differences.

It has been shown that the study of movements requires the observation of a definite number of parts of the body, and a definite number of loci of brain-tissue. In considering growth of the parts

of the body, organs, or tissues, the number of parts
or units is practically indefinite.

Movement is in many cases considered a reflex
act. Growth is, I suppose, never spoken of as a reflex
action. Growth is not often seen to occur as the
direct result of a temporary impression on a sensory
surface. A movement is said to be reflex when it
is excited again and again by a similar impression
upon the same organ of sense, or part of the sensory
surface of the body, and also when the movement
follows quickly upon that impression.

Movements in man are often considered as signs
of intellectuality ; acts of growth are never said to
be signs of mind or intellect in the subject growing.
Among the many beautiful adaptations of the parts
of plants we shall presently quote some which bring
about cross-fertilization. Now, such adaptations of
growth, such useful mechanism and action, are not
considered signs of intelligence in the plant, but
signs of some intelligence outside it.

The movements of an insect in visiting the flower,
and carrying off the honey, are considered signs of
its instinct or intelligence, the intelligence being
considered as a quality resident in the insect.

Then, again, referring to plant life, no amount of
adapted and useful movement in plants is con-

sidered to indicate intelligence in them, not even
the fanning movements of the *Desmodium gyrans*,
the rapid folding of the leaves of the *Mimosa pudica*,
or the successive folding of the tentacles of the leaf
of *Drosera rotundifolia* (see fig.7), are considered as
signs of intelligence, because the subject is a plant !

The results of growth are material changes, and
as such are permanent records of action, capable of
preservation.   Movements do not necessarily leave
any permanent record in the subject of such kind
as can readily be seen and preserved.

In speaking of movements as signs of the muscle-
stimulating power of nerve-tissue and of growth, we
deal with two functions in the body.   We study
man, and speak of nutrition and action in his body,
and of his functions; as examples, take growth
indicated by weight and movement.   The outcome
of nutrition in the child will vary at different ages;
in the early years much nutrition is consumed in
making weight, in adult life much energy is consumed in movement.   In a child becoming choreic
the weight falls, and movement increases in
quantity.   Function, then, means the outcome of
internal action; in the case of the brain the principal
function is to send efferent stimuli to the muscles.
In speaking of kinetic function, we mean this muscle-

stimulating function by which movements are produced. We speak of kinetic function in contrast to trophic action, which signifies growth or other material, structural, or molecular change.

When studying movements we found the absolute importance of noting the parts in action. So with growth, the parts growing must be clearly identified; they are most conveniently defined in anatomical language. The part growing increases in bulk, or undergoes an alteration in shape, which may be described by referring to certain axes of the body, or by describing the length, breadth, thickness, or circumference of the part.

Looking at growth in the human body, the same enumeration of parts may be made as when speaking of movements: bilateral parts, the members, etc.; further, small units or loci of tissue may be the seat of the changes observed. In studying movements it was not necessary to make as much subdivision of the body as is sometimes necessary with regard to the processes of growth. As to movements, we considered the parts that can move separately, and the nerve-muscular apparatus for each movement. These parts are finite in number; the units of tissue capable of separate growth are infinite. The complete study of growth involves the observation of at least all the separate tissues and cells of the body. In this fact lies the great difference in the methods of studying kinetic and trophic action, and their attendant difficulties. If we believe in one series of laws of nature for both kinetic and trophic action, we must expect to find the laws of movement to be those regulating growth.

We may observe the *time* of any movement, vital act or act of growth. We commonly speak of relations in time as the order of growth; in describing the processes of development, the order of development is most important.

The time of action of the muscle-stimulating function of the brain-centres has been shown to be regulated by such forces as light, sound, mechanical impact, or irritation. It is, then, desirable to inquire how far the time of acts of growth, or other vital acts, is regulated by such forces.

We more commonly study the time of kinetic action than that of trophic acts, because the processes of growth are slower, and observation must be continued for a longer period to see any change, than is the case in observing movement.

Visible movement indicates its own time to the observer, but does not necessarily leave any per-

E

manent record; growth is a process which usually does not impress the observer with the time of its acts, but leaves a result in visible form.

A visible effect of movement in man is to produce some fresh posture, or condition of temporary equilibrium.

With regard to observation of the effects of forces upon the subject. In the case of growth a much longer interval usually elapses between the stimulation of the subject and the sequential effect than that which takes place in a reflex movement. Hence more instances of growth are called "spontaneous," than in the case of movement.

Proceeding to apply the principles above stated, we may illustrate their application to various facts and observations.

To begin with examples of plant life, the time or order of events is marked as an important element in the following illustrations. In many insect-fertilized flowers cross-fertilization results from an arrangement in the time of development of the organs. Flowers are called proterogynous, when the stigmas are protruded and in receptive condition before the anthers have matured their pollen. On the other hand, they are called proterandrous when the anthers mature and discharge their

pollen before the stigma of that blossom is receptive of pollen. *Scrophularia* is a good instance of
proterogony in flowers fertilized by bees. The
following description of this flower is given by
Professor Asa Gray * :—" The flower is irregular,
and is approached from the front, the spreading
lower lobe being the landing-place. Fig. 13*b* repre-
sents a freshly opened blossom ; and Fig. 13*c* a sec-

Fig. 13.—Scrophularia.

tion of it. Only the style tipped with the stigma
is in view, leaning over the landing-place ; the still
closed anthers are ensconced below. The next day,
or a little later, all are as in Fig. 13*a*. The style, now
flabby, has fallen upon the front lobe, its stigma
dry, and no longer receptive; the now opening
anthers are brought upward and forward to the
position which the stigma occupied before. A

* " Structural Botany," p. 220.

honey-bee, taking nectar from the bottom of the
corolla, will be dusted with pollen from the later
flower, and on passing to one in the earlier state
will deposit some of it on its fresh stigma. Self-
fertilization here can hardly ever take place, and
only through some disturbance of the natural
course."

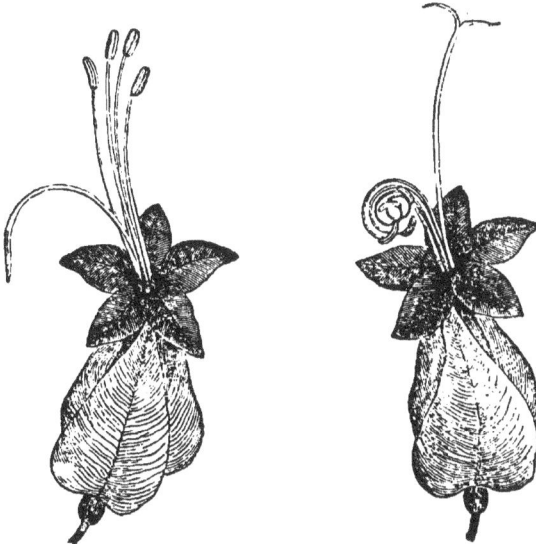

Fig. 14.—Clerodendron.

In *Clerodendron* a similar result follows, from the
anthers ripening before the pistil. The flower is
conspicuous from its colour. The long filiform
filaments and style project when the corolla opens ;
the stamens remain straight, but the style proceeds

to curve downward and backward, as in Fig. 14. The anthers are now discharging pollen; the stigmas are immature and closed. On the second day the anthers are dead and the filaments recurved and rolled up spirally, while the style has taken the place of the filaments, and the two stigmas, now separate and receptive, are in the very position occupied by the anthers the previous day. An insect flying from one flower to another must transport pollen from one to the stigma of the other.

These examples serve to illustrate the importance of the relations in time among acts of growth.

The time of acts of growth is an attribute bestowing special characters of its own on the subject observed: the time of commencement and the order of dentition; the time of growth of hair on the scalp, face, and pubis, are characters of the process being normal or abnormal. The duration of pregnancy is fixed, and the stage of development at each period of time has been fairly determined. Variations from the normal may be due to altered times of acts of growth in the life series. Periodical growth, with intervals of rest, are seen in the vegetable world accompanying the winter and summer seasons. In birds and animals the epidermal appendages grow at certain seasons, and are cast off at other times.

The second intrinsic attribute of growth is its *quantity*. In observing and describing movements as signs of vital action, it appeared that the time of action was the intrinsic attribute most easily observed, and it seemed to be the most important. In studying the motor-action of nerve-centres, we found that it was a point of much interest and importance to describe the balance of action in parts of the body, or the ratio of action in the different subjects considered. So now, in considering growth, we shall find that many phenomena may be described, and receive some kind of explanation, by noting the ratios of the quantities of growth in the parts compared. In attempting to describe growth in specimens, which are its results, we observe the outcome of action, the quantities of the growth of parts being more obvious than the times of growth. Looking at a specimen, as a heart, we have no clear indication of the results of the time of growth ; all that we observe are ratios of quantities of growth of its component parts, producing the structures and form of the organ.

This is analogous to the characters of motor-action seen when we look at a posture of the body ;

the characters of the posture are due to the ratios of action in the nerve-muscular apparatus.

These considerations lead us to study those relations as to quantity of growth in different parts, which impart special characters of their own to the subject under observation. Such observation is a mode of study analogous to the observation and comparison of postures of the human body, which are due to the ratios of kinetic or motor action, as growth is to what we call trophic action. This suggests the hypothesis that the forces which lead to certain postures in man, may be those which control growth in the body. The principles and modes of procedure found useful in dealing with postures will help us in deciding into what parts we may primarily divide the human body in studying proportional growth. Hence we compare the following parts as to their ratios of growth : bilateral parts; the members; the parts of the members; the large parts in contrast with the small parts, etc.

In describing the results of development in a living being, it is often convenient to do so in terms implying quantity of growth; examples may be taken from among vegetable and animal specimens.

The mode of development of buds will serve as

a good illustration of both ratios of growth and the times of growth in all parts concerned.

The rudiment of a leaf commences as a mass of cells growing outward from the stem; the rudimentary leaves are developed as plates of cellular tissue, such plates of tissue are formed quickly one below another, they envelope the shoot, and growing more quickly than it, they envelope it, and form a bud. This bud-formation depends upon the more rapid growth of the outer or under. surface of the leaves in their young state, by which they become concave on the inner (afterwards the upper) side, and pressed upwards to the stem. When perfectly developed, by the latest extension of their tissue, the leaves turn outwards in the order of their age, and thus escape from their position in the bud.

The wild carrot differs from the cultivated variety in the size and weight of the root. If we make a transverse section of the root it becomes obvious that the ratio of parenchyma to the fibro-vascular bundles is different in the two varieties, the cellular tissues being much more abundant in the cultivated variety. Professor Trail informs me that, as a sequence of this, the cultivated root is much more liable to disease from attack by fungus, while fluids from the soil have to penetrate through a thicker layer of cellular tissue to the vascular bundles, than in the case of the wild plant. The altered proportional growth of parenchyma and vascular bundles may be advantageous to the gardener but not to the plant.

In the growth of many fruits the ratio of growth of the seed and the pericarp, is such that the capsule enlarges rapidly, and forms a chamber surrounding the seeds, within which they lie in a cavity containing air. This is marked in the fruit of some of the acacias.

It is common to find in the leaves of plants that their general characters are determined by the ratio of their component tissues, the parenchyma and the fibro-vascular system. In the leaves of the cactus there is much cellular tissue, hiding the veins of the leaf, and giving it a succulent character. Contrasting such a leaf with the holly, the hard and prickly leaf is seen to derive its character from the small amount of cellular tissue in relation to the fibro-vascular.

There is a normal proportional growth for each member and part of the body. These ratios of growth define the normal form and development; mal-proportional growth is seen in cases of hypertrophy of the fingers.

The ratios which indicate proportional growth
may be taken between any of the parts of the body,
such as the ratio of the skull to the other parts of
the skeleton at various ages. Such observations
fall within the department of anthropometry.

We may investigate the proportional growth of
a tissue, such as fat, in different parts of the body.
This is found to vary much. A few examples may
be given. In children, fat is often absorbed from
the trunk and limbs, while the face remains plump.
In middle life, fat accumulates about the walls and
contents of the abdomen. The quantity of fat is
probably least variable about the palms of the
hands and the soles of the feet; perhaps this is
because these parts are more uniformly pressed.
In all these cases the proportional growth of fat in
bilateral parts is usually equal.

We sometimes find young leaves narrower and
proportionally longer than the older ones; the ratio
of the long axis and the transverse is altered as the
leaf grows older. This ratio is a sign of the age of
the leaf. In some leaves the proportional growth
is similar in the young and in the older members,
as in the nasturtium and ivy. In man the ratio
of girth and height varies with increasing age.

Unequal proportional growth of bilateral parts

is not uncommonly seen, and this is one of the common causes of curvatures in growing parts. A long bone may be divided into two halves by a plane passing down its axis; equal growth on either sides results in the bone growing straight; if the ratio of growth be altered, that side will necessarily become convex on which the growth is greatest.   Curvatures may result from unequal bilateral growth.

Looking at the development of the thorax in the fœtus, it is evident that the curvature of the ribs increases with growth in length.   There is probably but little muscular action upon the ribs, the diaphragm and the respiratory muscles not acting before birth.

The following quotations from Sachs express the laws of unequal bilateral growth in plants :—

" Even in multilateral erect stems, and vertically descending roots, growth does not always proceed equally, and with equal rapidity, on all sides of the longitudinal axis; it is much more common for first one side and then another of the organ to grow more rapidly than the rest, curvatures being thus convex, the convexity of which always indicates the side that is at the time growing most rapidly.   If another side then grows more rapidly it becomes convex, and the curvature changes its direction. Curvatures of this kind caused by the unequal growth of different sides of an organ may be called *Nutation*. They occur most commonly and evidently when growth is very rapid, and consequently in organs of considerable length, and are produced under the influence of a high temperature either in darkness or when the amount of light is very small.

"These movements of nutation of bilateral appendicular organs take place mostly in one plane, which coincides with the medial plane of the organ. As long as the organ grows most rapidly on the dorsal side, it may be termed, after De Vries, *hyponastic*; afterwards, when it grows most rapidly on the inner or upper side, *epinastic*." *

Curvatures must occur in a growing part if the bilateral growth be either unequal in quantity or asynchronous in time.

"With this bilateral organization is also usually connected a difference in the growth of the two unequal sides, which causes curvature, and hence changes in the position of the apex. The two unequal sides of bilateral organs must also be acted on differently by external agencies which

---

* Sachs, "Text-book of Botany," 1875, p. 765.

affect growth, such as light, gravitation, and pressure. We do not attempt here to solve the question of the causes which produce the bilateral structure in any particular case; it need only be shown incidentally that this structure of lateral organs is probably always brought about by internal causes, and is independent of the action of external circumstances. This is in general at once evident from the fact that the median plane of bilateral appendicular organs has always a perfectly definite geometrical relation to the axial structure which bears them, and that, moreover, in the dark and under the influence of slow rotation round a horizontal axis which eliminates the effect of gravitation, the bilateral structure and relation to the axis remain unchanged." *

After dealing with bilateral growth, we may observe the corresponding members of the two sides. They are generally similar, more or less, in their proportional growth; exceptions are worthy of note. In man we have almost universally "right-handedness;" the right limb, being the most used, is the larger in circumference. This fact seems to favour the view that growth is in part proportional to the quantity of nerve-stimulation of the part, for

* Sachs, *op. cit.*, p. 965.

the greater use of the muscles of the right limb indicates a greater amount of nerve-stimulus to it.

Proportional growth of members and their parts, as well as in the segments of the body, may be studied in the development of the fœtus. Such points might be noted as the position of the umbilicus; the relative length of the upper and lower extremities; the size of the head in relation to that of the body ; the relative curvature of parts ; the size and form of openings in the skeleton or in soft parts—each such fact being noted at successive stages of development.

We may now pass on to consider examples where the *antecedents of growth* have been determined. It will be readily granted that the forces which control quantities of growth are really the antecedents of the results of those quantities of growth. If we take the curved portion of the stem of a plant and show that the curvature is due to unequal bilateral growth, and that the unequal bilateral growth is sequential to the action of light upon the stem, then it may be said that light is the cause of curvature. The most important antecedents of growth are the same forces which we found capable of controlling motor function. The effects of light in controlling growth in different vegetable tissues and structures are well illustrated by the next example quoted from Sachs.

"It is remarkable that etiolation does not extend to the flowers. As long as sufficient quantities of assimilated material have been previously accumulated, or are produced by green leaves exposed to the light, flowers are developed even in continuous darkness which are of normal size, form, and colour, with perfect pollen and fertile ovules, ripening their fruits and producing seeds capable of germination. The calyx, however, which is ordinarily green, remains yellow or colourless. In order to observe this, it is only necessary to observe a stem of *Curcurbita* with several leaves, the main stem having been passed through a small hole into a dark box, the leaves which remain outside being exposed to as strong a light as possible. The bud develops in the dark a long colourless shoot with small yellow leaves and a number of flowers, which, except in the colour of the calyx, are in every respect normal. The extremely singular appearance of the abnormal shoots, with normal flowers, showing in a striking manner the difference in the influence of light on the growth of different organs of the same plant." *

* *Op. cit.*, p. 675.

Of all the forces concerned in producing proportional growth, it seems probable that Light exerts
the greatest influence.   The effect of light in checking the growth of internodes, and increasing the
growth of leaves, is a marked example.   Speaking
of the action of light on the growth of plants,
Sachs says, " It has already been stated that the
various parts of the flower grow as readily in permanent darkness as in light.   Most internodes, on
the contrary, grow more slowly when exposed to
light on all sides, and remain shorter when growing
in the darkness ; when the light reaches them from
one side only, they curve concavely towards the
source of light." *

Mechanical force or pressure may control proportional growth.   Growth in a bone may be determined as to its quantity by the action of muscles
attached to it.   The parts of a bone where strong
muscles are attached are much pulled upon by
them, and they tend to produce bony outgrowth at
such parts ; possibly such outgrowths are due to this
mechanical stimulus.   In a bone removed from a
muscular subject the points of insertion are well
marked.

Constant   pressure   upon   bone   may   cause   its

* *Op. cit.*, p. 752.

absorption, as has often been clearly demonstrated by specimens.

The ratio of the size of the foot to that of the body is reduced in a Chinese woman by mechanical pressure.

Friction and galvanism, applied to muscles that are in part cut off from their nerve-supply, aid their nutrition and growth, as may be seen in cases of facial palsy and infantile paralysis.

Pressure may aid the healing of ulcers of the leg.

Local irritation of the radicle of a seedling bean is followed by unequal bilateral growth and sequential curvature.*

Heat, or the temperature surrounding the growing plant or animal, is often a potent cause of the ratios of growth. The effects of climate may be quoted as causing variations of growth of the parts of many animals and plants. A sufficient temperature is essential to the germination of the embryo of seeds. Heat, as well as light, determines the quantities of chlorophyll and starch in plants. Warmth encourages repair in injured parts of the body.

Gravity affects the ratios of growth in the two halves of both the radicle and stem of plants, producing the phenomena known as geotropism and

* See Darwin on " Movements of Plants."

F

apogeotropism. If a seedling pea, about ten centi-
metres long, is laid in a horizontal position, the parts
already grown will retain this direction, but the
growing tip of the root will push downwards, the
growing tip of the stem upwards. The effect of
gravity is to produce unequal bilateral growth in
both stem and apex; in each case the convexity is
produced by the greater growth on that side. This
effect of gravity is removed by placing the plant on
a revolving table.

The interest attaching to the study of examples
of quantities of growth in different parts is to note
the *sequences* which follow therefrom. Let there
be a circular hole in an iron plate; if that plate be
made red-hot, it will enlarge in all directions; the
hole will retain its circular shape, but will be
larger than it was before. For similar reasons, in
the growth of bones in which uniform extension is
taking place in all directions, the foramina enlarge,
retaining their shape; this is seen to a certain
extent in the foramina of the pelvis.

In the skull of the infant the anterior fontanelle
becomes larger as growth occurs in the early stages,
owing to general enlargement of the bones of the
skull. This takes place up to about the ninth month
of life, when the fontanelle begins to close, owing to

altered ratios of growth in the parts of the skull, special quantities of growth occurring around the margins of the opening leading to its closure.

Among the sequences of proportional growth we may have increase of one tissue, such as the fat, markedly increasing the total weight of the animal.

If in the processes of nutrition, one tissue or one part grow in greater quantity than others, that tissue or part is said to be hypertrophied. Examples will be given in the last lecture.

Variation in proportional growth may, in some degree, alter the function of the subject. Increase of muscular tissue leads to greater motor-power; increase in the relative size of the salivary glands leads to more secretion.

It has been demonstrated that in cases of unequal bilateral growth curvatures may result, as in the growth of bones and in the stems of plants. We may quote one example, where the temporary swelling or enlargement of cells leads to important results. The stomata of plants are formed by two symmetrical semi-lunar cells, called guard-cells. The concave surfaces of these are to a certain extent mobile, and capable of being approximated one to the other. When they meet, from swelling of the guard-cells, the stoma is closed; when they elongate their concavity becomes more pronounced, and the stoma is opened. The use of the stoma to the plant is to allow gases to pass from the air to the cellular tissue under the epidermis, and *vice versâ*, also for the exit of aqueous vapour.

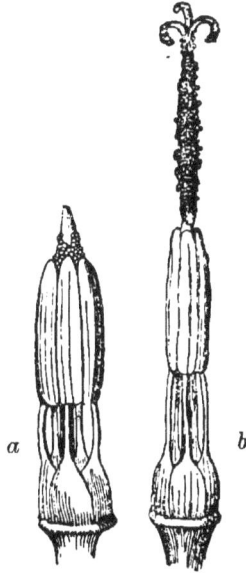

Fig. 15.—Composite flower.

a, Young flower, with pistil short. *b*, Older flower, with pistil and style protruding.

The sequences of proportional growth may be to produce cross-fertilization in flowers, as is well illustrated by the growth of a flower of the *Compositæ*. In the flowers of the Composite order a special arrangement is found with regard to the relative growth of the stamens and the pistil, which results in cross-fertilization of the flowers; that is, the flower, being hermaphrodite, is not self-fertilized. The stamens, being united by the margins of their anthers, form a tube, into which the pollen is discharged. The stamens grow to their full height and maturity before the pistil, which, during the early stage of the flower, is short and immature. When the anthers are ripe and have filled their tube with pollen, the style begins to grow, and passes up the tube formed by the anthers, pushing the pollen powder before it, which now accumulates in a heap at the top of the anther tube. Later, the continued growth of the style brings the stigma as the most prominent part of the flower; its lobes open and expose the receptive surface. An insect visiting the flower in its early condition, meets with a heap of pollen dust at the top of each floret, and thus dusts its abdominal surface; when the insect later on visits a flower in the later stage of growth, with the style protruded and expanded, it deposits some pollen from the former flower upon the prominent and receptive stigma. Thus cross-fertilization is effected.

The ratios of the growth of parts may bring

about asymmetry of form, malformation of parts, or
variations in the development, and consequently in
the use of parts. Examples are seen in descriptions
of short-horned and long-horned cattle, and in the
proportions of the beak of pigeons.

We considered examples where it was convenient
to describe actions in terms of movement, so as to
use terms which connote physical facts that can be
observed. We may now see how descriptions can
be given in terms of the outcome of trophic action.

Look at a germinating seed. The seed consists of
the embryo enclosed in its testa, or covering. As
long as it is dry no change occurs; when moistened
and kept at a sufficient temperature the process of
germination begins by the swelling and growth of
the embryo. But the testa does not grow; it may
stretch a little, but soon the increased size of the
embryo causes the seed-case to split. The splitting
of the testa is the outcome of unequal increase in
the size of the embryo and the seed-case. It thus
differs essentially from the growth of fruits.

We often speak of one subject growing similarly
to another, or of growth in two subjects being
similar. It often happens that the similarity of
growth is in part due to equal ratios of growth in
the two; in this sense, similarity of growth may

occur in two unlike subjects, if the ratios of growth
are equal in each. Male twins and female twins
often grow in equal ratios of similar parts in each.
Mr. Francis Galton has demonstrated this.   In
what characters may such similarity or resemblance consist ?   We discussed in Lecture I. the
characters which make one series of movements
similar to another, and found it to depend on
identity or similarity of the subjects considered,
and in identity of the attributes of the movements
or outcome of its action, *i.e.* in their time and
quantity of action. We found that movements in
two men may be said to be similar when they
occur in the corresponding parts of each, and are
alike in their time and quantity. In like manner
acts of growth are sometimes said to be similar
when they occur in the corresponding parts of two
or more subjects at the same time and in the same
quantity. In two pea seeds, placed under similar
favourable circumstances of moisture and temperature, the time and the quantity of growth will
probably be identical, the root and stem of each
growing equally in successive periods of time.

Similarity of growth may be partial; in two
subjects the growth may be similar as to time, or
quantity, or kind. When two processes of growth

are different in kind they may be similar in time; the growth of leaves and of underground tubers may occur at the same time, and this may be the only point of similarity between the two processes of growth. If the subjects in which growth occurs are alike, and the acts of growth in each are similar in their time, quantity, and kind, then we may say that the growth is similar in each, and the results of growth at any given time will be alike.

After describing acts of growth, we naturally pass on to say something about their classification. Our object is to classify facts concerning the processes of growth according to such intrinsic attributes as are observable, and to seek the causation of the attributes observed. Having classed a number of facts as examples of unequal bilateral growth, we may seek to determine the causation of unequal bilateral growth. In pursuing such methods of inquiry, we follow methods similar to those which aided us in studying the motor functions of the brain. Thus we have lines of thought to guide observations. It is possible, and I think highly desirable, to arrange a museum, in which specimens are classified according to the attributes of their growth, with sub-classes, in which the specimens shall illustrate the attributes in relation to the antecedent forces that stimulated their growth and produced that special ratio of the attributes.*   It seems to me that this was in part the principle followed by John Hunter in his physiological museum, which we possess this day.

Mill says,† "there is no analogy, however faint, which may not be of the utmost value in suggesting experiments or observations that may lead to more positive conclusions."

In making an analogy we can only compare things or acts as to the characters or attributes which they possess in common.   We may compare like functions as to their time and quantity ; it is only as to the time of action that we can compare unlike functions; we do not directly compare the material things, but the attributes of their functions.

Compare growth in the body of a child and in a leaf.   There are obvious differences in the two subjects, in the pabulum supplied to the growing parts, and in the mode of its supply.   Further, there are but few and simple tissues in the leaf, while many and very complex tissues are found in the body of the child.   In the following points we have characters which enable us to make analogies.

* See Catalogue appended to this lecture.
† " Logic," vol. ii. p. 93.

The time of growth in the child and the leaf may be compared; they may grow synchronously or asynchronously. We may compare the ratios of growth in the child and the leaf. Taking the ratio of length and breadth in each case, the ratios may be identical or not, but the ratios can be compared, and their antecedents and sequents observed. The ratios of growth may also be estimated by the ratio of the weight of the child and the leaf, at the commencement and termination of a given period, and thus the proportional growth may be determined.

Comparing two children, or a child of five years with the same child when twelve years old, we may compare the ratios of height and width in each; or, again, we may compare the ratios of the amount of trophic and kinetic action in each; or the height with the weight in each.

When two things, or two actions, are measurable it is often useful to compare them as to quantity.

We cannot correctly make a ratio between quantities which are factors of unlike units; we may not mathematically compare quantities of force with quantities of material, because they have no common unit of measure. Still, we find it convenient to speak of the ratio of height to weight, the ratio of weight and movement, etc.; the form of a ratio is used as a mode of verbal expression.

In our desire to give some kind of explanation of facts in nature, we use processes of analogy. Acceptance of the hypothesis of evolution encourages the study of analogies. Comparative Anatomy, as a science, consists largely of ·Morphological analogies. The theory of Evolution has added an additional interest to the study of analogies among all living beings, for this theory involves the idea of a common susceptibility of all living things to the action of one force, or set of forces, thus forming one link of similarity among all living beings.

One purpose in making analogies between processes of growth is, that we may classify examples in the best manner for seeking the causation of such processes.

We find an analogy between postures and proportional growth, both being the outcome of quantities of vital action, and we find that both are apt to be abnormal in the same neurotic subject. This will be described in the next lecture.

Malproportional growth in the members and parts of the body, producing several coincident defects in development, is usually accompanied by defective development of the nerve-system. Probably here again the coincident defects are those of quantity of vital action.

Analogy is sometimes studied in examples that are called *Reversion*. I suppose that the term reversion simply connotes that the action, or the outcome of growth observed, is similar in its intrinsic attributes to others that have previously been observed, the action (movement) being similar in its time and quantity, the outcome of growth in its kind and its ratios. The reversion is a repetition, or reappearance of vital acts similar in their intrinsic attributes, and occurring in similar subjects.

In the first lecture we spoke of repetition of similar series of movements, we defined what was meant by that phrase, and also what was meant by similarity of movements. We are now speaking of the repetition of the acts of growth, or the reappearance of conditions of growth in an individual or in a species, such as is commonly called reversion.

Speaking of Reversion, C. Darwin says,* " When a child resembles either grandparent more closely than its immediate parents, our attention is not so

* " Domestication of Plants and Animals," vol. i. p. 28.

much arrested, though in truth the fact is highly
remarkable; but when the child resembles some
remote ancestor, or some distant member in a
collateral line—we must attribute the latter case
to the descent of all the members from a common
progenitor—we feel a just degree of astonishment."

A child may grow like his father, or grandfather,
or like what his parent was at the corresponding
age; the similarity may be complete in height,
proportions, form and weight, etc. If the boy be
so complete a repetition of his father's growth, it is
probable that his kinetic acts or movements may
be similar, thus making the reversion as complete
as possible. The reappearance of conditions, or
results of growth previously seen in ancestors is
common. There may be reappearances of normal
or abnormal parts; malformations are often repeated in lineal descendants.

We have endeavoured to show that physiological
facts may be described in terms implying relations
of quantity and time of action, and that the time
and quantity of trophic action are controlled by
physical forces. It remains to enunciate our proposition concerning acts of growth, analogous to
that put forward with regard to movement. "Every
vital act, or act of growth, requires among its ante-

cedents a supply of pabulum and stimulation by some force incident to the subject." This is an important proposition or hypothesis to put forward, and can only be proved by the collation of many facts, and for this purpose more time would be needed than that at my disposal. It seemed, however, worth while to enunciate the proposition, because a large number of the examples given and arguments brought forward point to its truth.

CATALOGUE OF SPECIMENS

Arranged to demonstrate the importance of the Times and Quantities of Growth in the Parts of Animals and Vegetables ; also to demonstrate the Action of Forces upon Living Things in determining the Time and the Quantity of Growth in their Parts.

Examples of Proportional Growth.

1. Crystals of similar form and various sizes. Repair of crystals.

2. Leaves of *Eucalyptus* ; the young leaves are long and narrow, the older ones are broader.

3. Seedling maple, showing the different form of the lotyledons and the subsequent leaves.

4. Symmetrical and oblique leaves.

5. A chestnut germinating. The bursting of the seed is owing to the unequal enlargement of the embryo and the testa.

6. Seedling beans (*Vicia faba*). The arched hypocotyl is due to epinasty, or greater growth upon the upper surface.

7. "Mus. Cat. : Osteology," No. 3; No. 14, on development of bones. In a fœtus the ribs become longer and more curved as growth and development occur ; this may be seen in the ribs and the femora.

8. Specimens showing distribution of fat.

Examples showing Relations of the Time of Growth.

9. Flowers of *Clerodendron*, showing the anthers ripe, while the pistil is immature.

10. Flowers of *Clerodendron* in a more advanced stage, showing the anthers effete, and the style protruding and receptive.

11. *Orchis maculata;* a young plant which has produced tubers, and no flowers.

12. An older plant, showing flowers developed, and tubers diminished in size.

13. A jawbone, dissected, to show the third molars as yet uncut.

13a. Senile atrophy of heart, kidneys, and brain, from the same patient.

Examples showing some Sequences of Proportional Growth.

14. Calvaria of children, from birth till twelve months old, showing the actual increase of size of the fontanelle during early growth, and later its gradual closure.

15. A series of pelves at successive ages, showing increase of size, and change of shape of foramina as growth proceeds.

16. Model of stoma of a leaf, showing that variation in the size of the guard-cells regulates the size of the opening.

17. Leaves of cactus, showing that the ratio of parenchyma and vascular tissue is such as to make the leaf succulent.

18. Leaves of holly. The hard character of the leaf is due to the small amount of parenchyma. "Mus. Cat. : Gen. Pathol.," 121.

19. Fruit of acacia, in various stages of development, showing air-cavity in the seed-case, owing to proportional growth of seeds and pericarp.

Examples of Proportional Growth controlled by Pressure or Mechanical Strain.

20. Part of the stem of a plant that has grown bent, owing to its having been strained constantly in one direction.

21. Leaf and petiole of a climbing plant (*Solanum jasminoides*); the petiole being greatly thickened from pressure of the twig around which it is coiled.

22. Head of femur, with trochanters and muscles attached to it. The portions of bone most pulled upon have grown the most prominent.

23. "Mus. Cat. : Gen. Pathol.," No. 1. Bladder hypertrophied from stricture ; the frequent extra pressure exerted by the bladder has been followed by increased growth of its muscular wall.

24. Hypertrophied heart without valvular disease, the kidneys being granular. Continuous high arterial tension has been followed by increased muscular growth.

25. Foot of Chinese woman, much distorted, atrophied, and hindered in growth by artificial compression. "Mus. Cat. : Gen. Pathol.," 30.

26. "Mus. Cat.: Gen. Pathol.," No. 31. Five dorsal vertebræ partially absorbed from pressure of an aneurism.

26a. "Mus. Cat. : Gen. Pathol.," No. 26. "A vertical section of the bones of a knee-joint atrophied in consequence of more than three years' inaction."

27. "Mus. Cat.: Gen. Pathol.," No. 27. "Stumps of a tibia and fibula long after amputation. Both bones are reduced in size." (Hunterian.)

27b. Heart from rickety chest ; white patch due to thickened epicardium.

Examples of Quantity of Growth controlled by Light.

27c. Two branches of the same plant, one grown in daylight, the other in a dark box. The branch grown in darkness has long internodes, small leaves, and perfect flowers.

28. A young geranium, showing bending of the stem towards the light ; the curvature is due to checking of growth on the side lighted.

29. A seedling plant, showing the root bent away from the source of light.

30. Crustaceans, that have lost their eyes from long residence in darkness.

30a. Similar animals, with perfect eyes, from a light locality.

Examples of Quantity of Growth controlled by Heat.

31. Fruit grown in hothouse.

32. Similar fruit grown out of doors.

Examples of Quantity of Growth controlled by Nerve-force.

33. "Mus. Cat. : Gen. Pathol.," 37. Muscles atrophied sequential to anterior polio-myelitis.

34. "Mus. Cat. : Gen. Pathol.," 24. Optic nerves and globes of eyes. One eye being destroyed and not impressed by light, its nerve atrophied from want of stimulation.

35. Eyeball lost from herpes of fifth nerve.

Examples of Quantity of Growth controlled by Gravity.

36. The young plant was placed horizontally, the stem has curved upwards, the root downwards, owing to greater growth of stem on under surface, and the root on its upper surface.

G

Examples of Quantity of Growth determined by Vascular Supply.

37. Fruit from branches of apple-tree that have been bent down, causing greater supply of sap to fruit. On these branches the fruit is enlarged.

38. "Mus. Cat.: Gen. Pathol.," 108. Spur transplanted to comb of cock, showing overgrowth from increased vascular supply. (Hunterian.)

Examples where special Quantities of Growth in parts produce the Pathological Character.

39. Rachitic bones ; curvatures being exaggerations of the normal curves, probably due to unequal bilateral growth.

40. Rachitic bones ; curvature being due to pressure from weight of body.

41. "Mus. Cat. : Gen. Pathol.," 20. Atrophied heart, otherwise quite healthy. From a man aged sixty-five, who died of cancer of the pylorus.

Examples where the special Quantities of Growth in various parts form the Teratological Character of the Specimen.

42. Cleft palate, defective heart, the skin and ears being malformed ; also brain showing defective convolutions. All from the same patient.

43. "Mus. Cat. : Teratology," 321. "The skeleton of a full-time hydrocephalic foetus, with extreme shortening of the limbs."

43a. "Mus. Cat. : Teratology," 371. A chick at about the tenth day of incubation, in which the left eye has suffered arrest at a very early period. The right eye is abnormally large, and the lower beak deviates to the left, so that it crosses the lower one.

44. "Mus. Cat. : Teratology," 317. The hand of a

human infant at birth, with considerable shortening of the fingers. The ungual phalanges and the nails are much smaller than is normal. The thumb is short and thick, but its terminal phalanx is not abnormally defective. The mother had a similar malformation.

45. Hypertrophy of fingers. " Mus. Cat. : Gen. Pathol.," 14. " A wax model of a human hand, with the index and middle fingers congenitally hypertrophied."

Examples where the special Quantities of Growth in various parts form the signs of Evolution in the Species.

46. A wild carrot, showing smallness of the root, and the small quantity of parenchyma in relation to the fibro-vascular system.

47. A cultivated carrot, showing the large root, and the large quantity of parenchyma in relation to the fibro-vascular system.

48. A cultivated carrot, showing fungoid disease in the excessive parenchymatous tissue.

49. A crab-apple, showing small quantity of parenchyma in relation to weight of seed.

50. A cultivated apple, showing increase of parenchyma in relation to weight of seed in the cultivated variety.

51. Buds of chestnut, showing the different curves of bud-scales, and young leaves at different stages of development.

52. Leaves of nasturtium, showing that the form, or proportional development, is maintained as growth proceeds.

53. Leaves of ivy, showing similar form at various ages.

54. Flowers of rhododendron ; showing metamorphosis of petals into stamens.

55. Flowers of water-lily, showing metamorphosis of petals into stamens.

56. Coloured flowers, showing more irregularity of form of corolla than other less highly coloured and less conspicuous flowers.

57. Skulls and beaks of pigeons, showing proportional growth varying according to the species.

58. Casts and postures of the human hand.

Notes refer to the Catalogues of the Museum of the Royal College of Surgeons.

LECTURE III.

THE STUDY OF PATHOLOGY.

LECTURE III.

THE STUDY OF PATHOLOGY.

THE student of *post mortems*, and the pathological histologist, have done very much to advance our knowledge concerning disease as seen in its ultimate effects, while the experimentalist has laboured with some success to demonstrate its modes of origin. Still much remains to be done in observing the signs of pathological processes during life, and we need further guides as to useful methods of thought and research in making our inquiries.

We propose to study pathological facts as we studied movements, noting the seat of change, its time, its quantity, and its kind, as attributes of the process which we attempt to describe. Then we search for the antecedents of this fact, inquiring as to the supply of pabulum to the part, the pabulum itself, and the conditions of its supply, also the forces or stimuli acting, or that have acted, upon the subject or its progenitors. Having noted the antecedents of the fact, we note its sequents. By pursuing such methods of inquiry, methods suggested by the analysis of movements, I think that new knowledge may be gained.

In studying any fact in physiology or pathology, as in studying any scientific problem, we must first describe it in exact physical terms such as connote what may be observed. In giving descriptions for scientific purposes it is advisable to avoid using metaphysical terms, as also to avoid speaking of pathological states as special entities. Let us, then, seek to describe pathological states in terms connoting facts capable of observation. We will not say that "the bone grows curved," but that "the bone grows with unequal bilateral growth;" we describe the process observed, not merely its results, and then we are led to seek the antecedents and sequents of unequal bilateral growth.

We propose to give descriptions of facts according to this system, then to describe antecedent and sequential conditions; that will be to seek for an explanation of the facts in a methodical manner. The work in hand is then divided into two parts— to describe facts, and to seek their explanation in the antecedents and sequents of these facts. Following these modes of procedure it will soon appear that in many cases the incident forces which act, or have acted, on the material subject appear to be important agents in bringing about the conditions observed. The importance of pabulum and its supply in bringing about pathological processes has often been insisted on; I would insist on the equal importance of the forces, or stimuli incident to the subject observed, as necessary factors in determining the action in it.

In the first two lectures evidence was advanced to show that some stimulus, or antecedent force, must act upon the subject in which any vital action is seen. It also appeared necessary to define distinctly the exact subject in which action was seen. We should not be content to say we saw a man move; we must say what parts of the man we saw moving. For the purpose of studying movements in man, we may divide his body into all the parts that can move separately; his nerve-muscular apparatus is divided into all the parts that can act separately. Conditions of pathological growth must be studied in a like manner.

In the first lecture our work was partly analytical, partly synthetical. It was analytical when we

looked at all the movements of a man and demonstrated that the characters of series of movements
depend upon the time and the quantity of the action
of each component movement. It was also shown
by analysis that the outcome of a series of movements may be a relation to the surroundings. Our
work was in part synthetical as when we arranged
classes of movements, such as synchronous movements, asynchronous movements, and states indicated
by ratios of action expressed by postures. Again,
analysis furnished us with general principles, which
enabled us to proceed with the synthetical work of
classification; many facts were thus classified in the
second lecture. We now start with certain principles
of analysis, and classes to be filled in. By thus
rearranging some well-known pathological facts,
new relations and the antecedents of facts may
be demonstrated, and important analogies may be
shown.

There are two great classes of Pathological facts :
those indicated by Movements (conveniently called
Kinetic), and those indicated by acts of Growth
(conveniently called Trophic).

Facts demonstrated by movements are seen in the
living subject; acts of growth are also signs of life.
The results of movements are postures, which may

be represented by casts or drawings; the results of growth, normal or abnormal, may be preserved in the museum.

If the pathological fact to be studied is kinetic, either active or in a state of equilibrium, we analyse it as we analysed movements and postures, and it will then appear that in many cases both healthy and pathological characters often depend upon the antecedents, upon the surroundings, or upon the sequents, rather than upon the intrinsic attributes of the action. Similar relations between normal and abnormal acts of growth will be demonstrated.

It appeared, in studying kinetic and trophic acts, that there are some practical differences in the methods of considering them. In studying a specimen we commence by giving an anatomical description, and we usually find it necessary to speak of its various parts; we thus divide it into minor parts. In examining a specimen which is the result of growth, we describe it as we see it, and have no immediate concern with relations as to time, only with relations in quantity and kind. After giving such description, we may seek the antecedents and sequents of the acts of growth by which the specimen was formed.

In seeking to determine the antecedents of pathological processes, we must remember the two important factors of nutrition demonstrated in the first lecture : the supply of pabulum to the part nourished, and the stimulus aiding and controlling nutrition.

Pathological processes kinetic in outcome, are principally met with in the clinical study of the nerve-system. As examples, the following may be mentioned :—chorea, athetosis, shock, coma, and all mental states. The principles of study recommended may be applied to examples of hypertrophy and atrophy : variations in proportional growth, bilateral, asymmetrical, etc. They may be used to explain the curvatures of bones.

There is a very prevalent, and apparently well-founded, opinion among biologists and pathologists, that the modes of action termed reversion are potent factors in the production of many of the conditions termed pathological. This idea has principally guided investigations concerning the action of the nerve-system and the origin of morbid growths. Such views may be illustrated by the study of the pathology of chorea, a condition indicated by movements.

I shall not attempt to prove any view, but put forward an hypothesis for the purpose of illustrating what I believe to be advantages derived from the definite observation and study of movements. The theory may be advanced that "Chorea is a condition of the brain analogous to that found in healthy infancy, such brain-state occurring at a period of life when the force generated by nutrition is greater than in infancy."

It is useless to speak of chorea as a condition of reversion unless we state in exact terms what points are to be observed to prove or disprove the proposition. A reversion has been defined as a repetition of a series of movements or trophic acts, similar to a series previously existent; the criteria indicating similarity have also been defined.

The child can never again become an infant, but it may become "infant-like." Its body is heavier and larger than that of the infant, its quantity of nutrition and its quantity of movement are greater. It is not suggested that chorea is merely a reversion of movements; it may be a reversion of modes of action in the nerve-centres.

In chorea the reversion is indicated by a repetition of the ratio of the attributes of the infantile state. The ratio of kinesis to weight is repeated, although of course the total quantities of both weight and movement are greatly in excess of

those in the infant. It is infant-like in the ratio
of kinesis to trophic action.

The condition chorea is due to an altered ratio of
functions; it is a reversion to the infantile ratio.

The problem is one concerned with the nerve-
centres, as to their kinetic, or muscle-stimulating
action. These centres are observed at two periods
of life : infancy and young childhood. An analogy
or comparison is made as to the quantities and time
of their action. Observation shows the combinations of movements in the young infant and in the
choreic child to be similar; what is meant by similarity of combinations of movements has been
defined. There is said to be a similarity as to the
parts, and their order of acting.

The similarity is thus partial; it is made as to
parts moving, and their order (time). Is this a fair
analogy ? Let us inquire further.

The quantity of movement depends upon brain
nutrition; the total quantity of nutrition in the
young child is much greater than in the infant.
The condition of brain as to its tendency to produce
movement must be taken as the ratio of its nutrition to the quantity of movement it produces.
Roughly we may take the relative weights of the
two ages as criteria of the ratio of nutrition; say,

the infant of fifteen pounds and the child of forty-five pounds. The nutrition, then, is three times as great in the child as in the infant; if the ratio of motion producing power and nutrition were maintained during development there would be three times as much movement in the child as in the infant. I believe this is not usually the case, but as development advances kinetic action is lessened in its ratio to other functions.

We say that in health the ratio of kinetic action to nutrition lessens as growth proceeds; if this ratio revert to that of infancy, there will be an amount of movement altogether abnormal. It seems to me that this reversion to the infantile ratio between the quantity of nutrition and movement is what we observe in chorea. It would occupy too much of your time to advance all the facts at hand in support of this hypothesis; my object is now only to show that the principles of study advanced make such hypothesis intelligible.

If chorea be such a reversion to an infantile state, what do we know concerning the causes of reversion? It has been shown by C. Darwin that circumstances which lower nutrition tend to produce reversions. It has been established that in chorea the body-weight falls; the loss of weight is not necessarily in proportion to the amount of movement, and often precedes the chorea. Increasing nutrition and weight by high-feeding is the most efficacious treatment; * cases in which the weight is low do not recover until it improves. Mitral regurgitation is often present; this leads to lessened and irregular supply of blood to the brain. Lastly, recovery is usually complete when the normal body weight is recovered.

The careful study of chorea, and the systematic observation of children for slight movements, and as to the postures which express the balance of action in the parts of each hemisphere of the brain, show the special liability of certain nerve-muscular areas to be affected by temporary or functional conditions. Thus, chorea may affect the whole of the area weakened by hemiplegia. This area may be affected on one or on both sides, it may be accompanied by aphasia. Further, the facial muscles, the tongue and hyoidean muscles, the pterygoids and masticatory muscles, the external ocular muscles, the spinal accessory nerves, may all or each be affected in chorea, and the same groups of muscles may be found twitching in a neurotic child.

The condition termed *coma* may be described in

* See Mr. Battam's paper in the *Lancet*, 1877.

terms indicating movements of parts, and absence
of movements or reflex movements. Standing by
the bedside of a patient we may note the general
absence of all but organic movements. Absence of
the following reflex-actions indicates that there is
complete coma :—The head is not turned towards a
source of light or sound; the word of command and

Fig. 16.—The Feeble Hand.

pressure on the chin are not followed by protrusion
of the tongue; no action follows upon our request.
The tone in the orbicular muscle of the eye is sufficient partially to close the lids. There may be
some movements of the limbs not controlled by
sights and sounds around, and subsultus tendinum
may be seen.

H

As to the condition of the eyes in coma, a description was given in the first lecture.

If in such a case the forearm be held up by the wrist-band of the shirt, the hand is allowed to fall free, and will probably assume the posture of "the feeble hand," flexed slightly at all the joints, with slight contraction of the metacarpus.

Many states of the nerve-system are known to us by physical signs described in terms of nerve-muscular action; such descriptions must indicate the part moving, and the time and quantity of action. This may be illustrated by making a comparative study of hemiplegia, with other brain conditions affecting special nerve-muscular areas. In hemiplegia we see lessened nerve-action in a certain muscular area. In the upper and lower extremity, the diminution of power is greatest in the digits and small parts, less in the muscles passing from the trunk to the limbs; in the face the muscles about the angle of the mouth are most weakened. The same area may be convulsed in hemispasm, and in hemichorea the same area is the seat of clonic spasms or twitches. In the comparative study of these conditions analogies may be made. The conditions resemble one another in the kind of action, all being motor phenomena;

they resemble one another in the nerve-muscular area affected; they differ as to quantities of action; motor-power is diminished in all the parts concerned in the hemiplegia, and increased in all the parts concerned in the hemispasm. Again, we find analogy between the two states in the comparison of quantities of action in the various parts. The decrease of quantity of motion in hemiplegia is proportional to its increase in hemispasm, in the small and large parts respectively. The motor signs indicating the two conditions are partially analagous as to the attribute time; the spasm first attacks the small parts, and these are the last to recover in hemiplegia.

Lessened nerve-muscular power does not always depend upon organic lesions. Fatigue and exhaustion of the brain-muscular power is very common. If both sides of the body are equally weakened, the man is said to be exhausted, and the condition is considered physiological. If, however, in a tired woman one side of the body, or one upper extremity only, be weak and powerless, there being no evidence of organic disease, then the state is often called Hysteria. It appears, then, that a bilateral diminution of nerve-force of short duration, as seen in fatigue, is considered by us as less pathological in character than a one-sided weakness. When examining patients as to their average postures and fine movements, it is very common to find the signs of unequal action on the two sides. Usually the left side of the body shows the weaker condition. This will be illustrated in the description to be given of a nervous child.

In studying movements and postures, as signs of the condition of the patient, the following suggestions may be useful. Observe whether movements be mostly in the small muscles or in large muscles. Thus, in observing the outstretched hand, movements of the finger or of a single digit may be seen; note if the movement be lateral in direction, due to action of the small interossei, or flexor and extensor, the result of action in the larger flexors and extensors in the forearm.

Any departure from symmetry of action, *i.e.* symmetry in the parts acting, is always important. The hand postures often differ on the two sides; all head postures and movements, except flexion and extension, are due to unequal action on the two sides.

The face is a muscular region where action is often seen in different quantities in the upper, middle, and lower zones; but the action is usually similar in quantity in the corresponding parts of the two sides. The only common asymmetrical expressions of the face are winking, smiling, and one-sided grinning.

I believe that in many children the early stage of lateral curvature of the spine is due to weakness of the central nerve-system, leading to asymmetry of posture of the spine, and that such cases should be treated accordingly.

We may now give a description of a *Nervous child* in terms of nerve-muscular action and ratios of growth. A typical case may be found among children who sleep badly, talk at night, grind their teeth, emaciate without disease of organs. Such children are apt to be irritable and passionate, and to suffer from headaches and hacking cough without lung disease.

Let such a child stand up, and observe it.

As to its conditions of growth, defects of proportional growth are commonly seen. The congenital form of the bones, the make of the skin, the form of the features may all be good. The height of the child in relation to circumference, and to its weight, is defective; the child is too tall and too thin; either fat or muscle may be defective in quantity. The emaciation may be equally dis-

tributed; often it is less in the face than in the
limbs and trunk. Glisson pointed out this fact in
his description of " Rickets " (see p. 117).

Now, as to the motor signs indicating the state of
the nerve-system. Let the hands be held out, with
the palms downwards, and the fingers separated.
The left upper extremity is often at a lower level
than the right; "the nervous hand" is seen on
either side (see Fig. 10, p. 24), perhaps more marked
on the left. There may be finger-twitching, separate digits moving in flexion, and extension, or
laterally in adductor and abductor movements. The
spine is arched too forward in the lumbar region,
often with inequality in the level of the shoulders,
and slight lateral curvature. The face, as a whole,
is usually too immobile, although there may be some
over-action of the muscles widening the mouth on
one or on both sides. The tongue when protruded
is too mobile. The eyes move, mostly in the horizontal direction, their movements not being fully
controlled by the sights and sounds of objects
around, except under strong stimulation. The head
is sometimes partially flexed, with inclination and
slight rotation towards the same side.

Some of the teeth are usually found ground at
their tips ; this is most commonly the case with the

canines. This grinding action is produced by the masticatory muscles during sleep, and is owing to irritation of the fifth pair of cranial nerves. We may here call to mind the fact that the sensory division of the fifth nerve is distributed to parts inside the skull as well as to those outside it.

I have given an analysis of fifty-eight cases of the neurotic condition in children in a paper published in the *British Medical Journal*, December 6, 1879.

Other examples might be given of defects in ratios of growth in the body coinciding with defects in the nerve-system; such coincidences are very common in idiots.

We now pass on to speak of unusual, or abnormal, conditions of growth.

In the last lecture it was demonstrated that ratios of growth have much to do with the character and the results of acts of growth. We are thus led to seek examples where the pathological character of growth is due to the ratios of action. We saw that many forces can regulate proportional growth; they will be seen in some cases as the antecedents of pathological processes.

Proportional growth is an element of great importance in many pathological conditions, and constitutes an important element in the description of many pathological processes. It will be found that many facts may be gathered under this head. In this class we shall consider examples of hypertrophy and atrophy, symmetrical and asymmetrical growth, bilateral growth, and the ratios of growth in the members, and the parts of the members of the body; we shall also include bone curvatures and malformations when clearly examples of mal-proportional growth.

The terms *Hypertrophy* and *Atrophy* indicate relations of the quantities of growth to the normal, the former term being used when the quantity is greater than the normal, the latter when it is less than normal. Hypertrophy and atrophy may be taken as indicated by ratios of growth to quantities of normal growth, not to quantities of growth in other parts of the same subject.

Our principal interest is as to the causes of hypertrophy and atrophy. Let us recall the antecedents of increase and decrease of movements. We found them to be principally the supply of pabulum, and stimulation by some form of force. The blood supply and stimulation are the usual determinable causes of hypertrophy, as is illustrated by the following examples:—" Mus. Cat. : Gen. Pathol.," Ed. 1882, No. 108 : " The head of a cock dried. A spur transplanted into the comb has grown into a kind of horn, which is about three-fourths of an inch thick at its base, about six inches long, curved forwards and to the left side, and deeply bifid at its extremity. Hunterian." The extra supply of blood received after transplantation, appears to have been followed by hypertrophy. Other examples might be given from the same source.

We shall next proceed to give examples where the quantity of growth, as influenced by mechanical irritation or pressure, is above or below the normal, leading to hypertrophy or atrophy. Look at the specimen of a Chinese foot:—"Mus. Cat.: Gen. Pathol.," No. 30: "The left foot of a young Chinese female, much distorted, atrophied, and hindered in growth by artificial compression." Other examples might be given of atrophy where pressure of tumours, aneurisms, etc., have caused absorption of bone.

Hypertrophy, or increase in the quantity of growth, resulting from intermittent pressure is seen in the case of corns and bunions, and also in the condition of the left ventricle of the heart when the arterial tension has long been high, as in cases of arterio-capillary fibrosis accompanying granular contracted kidneys.

A good example of hypertrophy from pressure is

given by C. Darwin, *in the *Solanum jasminoides*:
" The flexible petiole of a half or a quarter-grown
leaf which has clasped an object for three or four
days increases much in thickness, and after several
weeks becomes so wonderfully hard and rigid, that
it can hardly be removed from its support. On

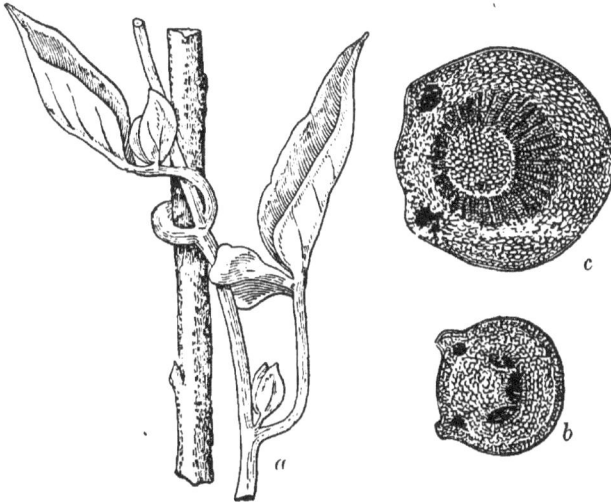

Fig. 17.—Solanum jasminoides.

b Transverse section of leaf-stalk not pressed by twig.
c Transverse section of young leaf-stalk which has clasped a
support.

comparing a thin transverse slice of such a petiole
with one from an older leaf growing close beneath,
which has not clasped anything, its diameter was

* " Climbing Plants," p. 73.

found to be fully doubled, and its structure greatly changed."

We shall refer to heat, and other modes of force, affecting the quantities of growth further on.

Nerve-force, whatever that may be, is a potent regulator of quantities of growth and nutrition.

When a limb is cut off from the nervé-centres which supply it with nerve-force, its muscles become atrophied, the trophic action in them is diminished, and the bones may in time become shorter than those of the healthy limb. Look at this specimen from a case of anterior polio-myelitis. ("Mus. Cat. : Gen. Pathol.," No. 37.)

" A portion of the leg and thigh of a young man who had suffered with infantile paralysis of the limb. The back of the limb has been dissected, and shows the thread-like atrophy of the nerves, the complete fatty degeneration of the muscles, and the unaltered texture of the tendons. In the transverse section hanging below, the gastrocnemius and soleus appear reduced to masses of fat. In this section, also, is well shown an accumulation of healthy-looking subcutaneous fat, in a layer more than half an inch in thickness, and constituting nearly half the size of this part of the leg. The limb was completely useless."

The loss of nerve stimulus, from a lesion causing hemiplegia, if it occur in early life, may be followed by deficient growth in the limbs paralyzed.

C. P., aged twelve years, became hemiplegic, probably from syphilitic disease of the brain five years before I saw him. There were athetoid movements in the paralyzed arm, and there was considerable rigidity. The paralyzed arm was somewhat atrophied, in comparison to the other side, and the forearm was half an inch shorter than the other.

A case may now be given in which a lesion of twenty years' standing, affecting the right fifth nerve, has been followed by atrophy of the masticatory muscles, and atrophy of fat of the orbit.* J. C. came under my care for an affection of the lung. Twenty years previously he began to notice the signs of the condition now to be described. On the right side of his face the temporal and masseter muscles were greatly wasted, the pterygoids also seemed weak ; the fat of the right orbit was greatly diminished in quantity, leading to sinking of that eye-ball into the orbit. The eye was healthy, and sensation in the face perfect.

It has been said that atrophy follows the disuse

* See *Lancet*, January 7, 1882.

of parts of the body, and that hypertrophy results when they are often and strongly used. I would argue that muscles often become atrophied when they receive too little stimulus of nerve-force, and that when they are well supplied with good blood, and are often strongly stimulated by incident nerve-currents, then their growth is increased thereby.

It is necessary to inquire in this place, What is meant by the use and disuse of parts? This phrase is usually employed as indicating the so-called voluntary use, or absence of the voluntary use, of muscles. The general statement may, I think, be used in a wider sense. Voluntary use is followed by increased growth; but we want to bring within the same category of facts examples where the organs of special sense, and other parts, grow larger when much used. In contrast with examples of excessive growth from much use, we may place others where atrophy follows disuse. It will be argued that all such cases are really examples of regulating of the quantity of growth by various stimuli.

Let us examine cases where atrophy of muscles follows disease of the nerve-centres. In anterior polio-myelitis the muscles waste from want of innervation, not from want of use; if the limb be stimulated by warmth, friction, and electricity, its nutrition may in part be maintained. In hysterical paralysis the limb is not used, but its nutrition is maintained by the nerve-centres.   Mr. C. Darwin gives many examples of loss of parts in a species from disuse.   I will quote but one of them.

"It is well known that several animals belonging to the most different classes which inhabit the caves of Cariniola and Kentucky are blind.   In some of the crabs the foot-stalk for the eye remains, though the eye is gone.   As it is difficult to imagine that eyes, though useless, could be in any way injurious to animals living in darkness, their loss may be attributed to disuse.   In one of the blind animals, namely, the Cave-rat (*Neotoma*), two of which were captured by Professor Silliman, at about half a mile distant from the mouth of the cave, and therefore not in the profoundest depth, the eyes were lustrous and of large size ; and these animals, as I am informed by Professor Silliman, after having been exposed for about a month to a graduated light, acquired a dim perception of objects." *

In this case it appears that the long exclusion of light from the eyes of a species is followed by loss of the eyes.

It appears that " voluntary action " leads to more

* " Origin of Species," p. 110, *et seq.*

growth than "spontaneous action." Chorea, athetosis, senile tremor, do not lead to hypertrophy, such movements are probably produced by intrinsic stimuli only, for they are not co-ordinated by forces outside the body. Athletic exercises and the movements of the blacksmith's arm are, of course, produced by his nerve-muscular actions; but these are stimulated and co-ordinated by the sight and sound of things around. Probably movements that are stimulated from without are more strongly stimulated than those that are spontaneous or due to intrinsic forces. Hence processes of training are used for the purpose of increasing muscular power.

If this view be accepted as a working hypothesis, we shall not say that voluntary use of parts is followed by hypertrophy, but that growth is increased by the amount of stimulation of the parts concerned, whether such parts be muscles or organs of sense. Further, we do not speak of "volition" as the antecedent of a physical fact; the physiologist cannot correlate "volition," with other forms of force producing physical changes.

We may next proceed to demonstrate that many unusual or pathological cases of growth may be due to the ratios or quantities of growth. Some miscellaneous examples will be quoted, and then we shall study in some detail the cases of Rachitic and Teratological specimens.

From among the group of changes in the growth of the body which make up the signs of old age, we may select the following examples of altered proportional growth. Those to be first mentioned are almost normal to the time of life; the latter are clearly pathological, as tending to destroy the life of the individual. At the approach of old age we see diminished growth; atrophy of the hair of the scalp, the skin, the muscles, and bones, while the fat may be either absorbed, or may accumulate about the abdomen, the heart, etc. Coincident with such changes the lungs and the heart may lose weight. The brain also often atrophies. At the same time increased quantities of growth may occur in the prostate, and in the hair bulbs of the chin of the female, while in some cases overgrowth is further seen in the development of epithelioma, sarcoma, and cancer. It seems somewhat vague to speak of senile overgrowth; let us inquire with what conditions we may compare it. If we compare the processes seen in the examples quoted, they agree in being altered quantities of growth occurring in ratios different from those found in early life. Probably they are less controlled by the forces surrounding the subject; that is to say, the subject is less impressionable in old age.

In Lecture II., when speaking of proportional growth in bones, it was shown that the form of the foramina is nearly maintained as growth proceeds. In the skull of the infant the ratios of growth are not constant; as a result of this the fontanelle is closed. The ventral openings for the spermatic cord close in the course of development by increased growth of the tissues around them occurring in a normal ratio, which differs from that of the earlier stages.

An analogous case is found in the process of closure of the auricular septum of the heart; when the foramen has remained patent, I prefer to say that the quantity of growth in the parts of the septum about the foramen was deficient, or below the normal, rather than to say that the opening remained patent, because this mode of description shows the natural allies of the phenomenon, and may help us to determine its causation.

Pathological conditions may be indicated by abnormal ratios of growth of the different tissues, the cellular tissue is in excess in many cases of the strumous diathesis. Fatty tumours, myomata, exostoses, are overgrowths of tissue normal in kind.

I

Excessive quantity of growth of one tissue, not abnormal in itself, may predispose to disease; the excessive quantity of parenchyma in the root of the cultivated carrot, as compared with that of the wild plant, predisposes it to attack by fungus diseases. These facts may add an interest to the study of the antecedents of proportional growth, and of normal and abnormal ratios of growth.

In the study of the condition of the bones in rickets, we shall find an application of the methods of study that have been put forward in these lectures. Our concern here is not with the general pathology of the condition known as rickets, but to see how the principles of inquiry which have been deduced from the study of movements are applicable to a wide field of observation. In the shafts of the long bones the curves are probably produced in two ways, by mechanical strain or pressure, and in other cases I think the curve is due to causes leading to unequal bilateral growth. The sharp curves sometimes seen in the clavicles, the bending of the ribs, and the forward curves of the femora of children who have walked, especially if they be asymmetrical, may perhaps be produced by strain and by pressure; while curves which are only exaggerations of the normal curves, and which are symmetrical, are probably due to unequal bilateral growth. These latter are often seen in children that have never walked, and whose muscles are very feeble ; further, in such cases the skull is often defective in form, being a bony growth but slightly acted on by muscles at the part most mis-shapen.

In favour of the hypothesis that intrinsic growth is a cause of some, at least, of the curvatures of rickety bones, we have the fact stated by Glisson, and now acknowledged by many, that the rickety bones are often hard, and not soft as has been stated. Mr. R. W. Parker, in his article on Rickets,* says " he has osteotomized many deformed rickety bones in children from three to ten years of age, and has always found the bones remarkably hard. More-over, in the *post-mortem* room, he has never seen bones *presenting the ordinary rickety curves* in a soft stage, nor have any such specimens ever been demonstrated at the Pathological Society. If the bones were really softened, it seems hardly probable that the resulting deformities would be so uniform as they actually are in practice, while the fact of their becoming straight without again softening, suggests some other explanation."

* " Heath's System of Surgery," p. 357.

I believe the opinion that growth is the cause of curvature of bones, is entertained by many; but the method of production of these curves is not clearly explained. It was the studies of which a sketch is given in these lectures, and particularly the consideration of the modes of growth in plants, which suggested the views here put forward as a working hypothesis. After my notes had been written I found that Glisson had arrived at much the same opinion.

The long bones present certain characteristic deformities and curvatures; these curvatures of abnormal growth are usually exaggerations of the normal curves. We are here concerned with the mode of production of these curves; no other tangible view than that of unequal bilateral growth has been put forward, and this hypothesis may serve to correlate the facts with other processes in nature. It appears that the ordinary rickety curves are the outcome of irregular ratios in the bilateral growth of the bones. Glisson's work on Rickets so clearly enunciates similar views, that it may be worth while to give quotations.

In flat bones, and in the pelvis, the foramina may be altered in size and form by the abnormal proportions of growth. When good nutrition in the child is restored, it tends to produce symmetry

, of growth, growth in due proportion in the parts of the skeleton. The proportional bilateral growth, and the ratios of growth in the various segments, are followed by normal curves and normal ratios of the epiphyses.

It seems to me probable that the defects in the pelvis are due to abnormal intrinsic proportions of growth rather than to pressure and mechanical causes, for not only are the axes and diameters abnormal, but also the foramina.

Extracts from Glisson on Rickets.

"Of how great moment the Alogotrophy, or unequal Nourishment of the Parts in this affect, we have already shewed; we shall here therefore prosecute those signs which in some great measure depend upon it, and we shall present them as if they were to be beheld at one View.

"First, there appeareth the unusual bigness of the Head, and the fulness and lively complexion of the Face, compared with other parts of the Body. . . .

"Secondly, the Fleshy parts, especially those which are full of Muscles beneath the Head, which we have listed among the first affected in the progress of the Disease, are dayly more and more worn away, made thin and lean. . . .

" Fourthly, some Bones wax crooked, especially the bones called the Shank-bone, and the Fibula. . . . Again, there is sometimes observed : a certain shortening of the Bones and a defective growth of them in respect of their longitude. . . .

" Finally, to this place also may belong a certain sticking out of the Bones of the Head, especially of the Bone of the forehead forwards. For it concerneth the common kind of viciated Figure and the Alogotrophy of the Bones. Yet this in the Bone of the Forehead doth evidently seem to depend upon the free nourishment of that Bone in his circumference where-with, it is coupled to the Bones of the fore part of the Head, and constitutes that seam called *Sutura Coronalis* which lieth in the foremost parts thereof. For herupon it must needs be thrust forwards. And indeed in that place it is plentifully nourished without any difficulty, because this Bone in Children is cartilagineous towards that Seam. And this also was preterpermitted above where we discoursed of the Organical faultiness, because we have but lately observed it. . . .

" Some may reply, That what hath hitherto been spoken doth concern the nourishment of the Bones in general, but that they yield not a reason of the protuberances in them.

"We deny it not; but seing that these faults of the Bones depend upon their unequal nourishment, as we have already proved, we supposed it would not be unprofitable to purpose some reason of their nourishment in general: now we draw nearer to the aforesaid swellings of the Bones. And we observe, secondly, that those tumors of 'the Bones are not of a different kind in respect of the other parts of the same Bone, but that they are parts altogether similary and of the like kind with the rest, and that they are not faulty in respect of the similary Constitution, but in respect only of their greatness and figure. . . .

"Fourthly, that the said swellings are produced by an unequal nourishment of the Bones, as by a more liberal nourishment of the swelling parts, and a sparing nourishment of the other parts of the same Bone. And these three last observations we have already abundantly proved where we treated of the affected parts. . . .

"Some ascribe this crookedness of the Bones, to the bending faculty of them ; for say they . . .

"But we cannot yeeld our ful assent in all respects to these Reasons. . . . We compare the Bones therefore, in which this crookedness useth to happen, to a Pillar, and not unaptly, seing, that

when they are erected, they resemble a Pillar ; and
from thence we deduce a demonstration that illustrates and makes the matter very plain. Let the
Pillar therefore consist of three stones A B C placed
over one another. We suppose it such an one as
is perpendicularly erected on every side, and of the
same height ; If therefore you shal fasten in a wedge
on the right side between the stones A B through

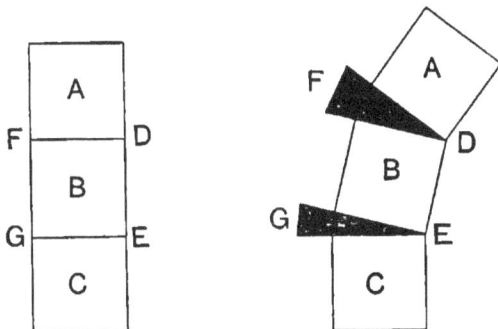

Fig. 18.

the line F D, the Head of the Pillar; Namely,
the stone A will of necessity be bended towards
D and wil make an Angle in D, and the height of
the Pillar on the right side, wil be higher than
on the left. In like manner if you drive in another
wedge thorow G E into the stones B C the pillar wil
be yet more bowed, and the angle will be made in
E. Now therfore the Pillar stands bent to the left
hand after this manner.

"We wil ad for a higher confirmation of this matter. An observation raised from the case of this crookedness of the Bones. The Quacks of our Country are wont to rub dayly the hollow, not the convex sides of the Bones, and that rubbing, say they, doth very much conduce to the cure, but this doth rather hinder it. But it is certain that rubbing doth powerfully summon the nutritive juice out of the Bloody mass into the part so rubbed, therefore if at any you rub that hollow part which is insufficiently nourished, it is no wonder if it do good, seing that thereby the aliment is more plentifully allured, and the heat of the part is also excited and augmented. . . .

"The common caus almost of al these recited affects seems to be an uneven or disproportionate nourishment or Alogotrophy of parts. Now this dependeth chiefly upon two causes in this affect. The first is, the unequal inherent Constitution of the parts irregularly nourished."

Defects in proportional growth are often seen in teratological specimens. Two have been selected from the museum.

No. 371 : " A chick, about tenth day of incubation, in which the left eye has suffered arrest at a very early period. The right eye is abnormally large, and the upper beak deviates to the left, so that it crosses the lower one." Here the over-development of the right side of the beak is the cause of the bending of its apex across the median line. The bending of the beak is due to unequal bilateral growth.

An analogous mechanical effect is seen in cases of enlargement of one side of the tongue from inflammatory swelling, the apex being thus displaced from the median line towards the smaller side, so when one side of the tongue is atrophied the apex points to that side.

No. 314: "Wax models of two human hands and a foot, with the index and middle fingers of the hands and the corresponding toes excessively hypertrophied." Here the abnormality is a mal-proportional growth, and it seems probable that this abnormal ratio in growth is due to some kind of stimulus transmitted from the parents to the fœtus, rather than to a special supply of blood to the overgrown parts. This seems to be rendered the more probable by an inspection of specimen No. 317, in the description of which it is stated that the mother had a similar malformation.

An interesting communication, having reference to twenty-seven individuals and five generations, is summarized as follows in the *Revue International des Science Medicale,* November, 1886 *:—First generation: Man born in the year 1752 had six toes on one foot. Second generation: A son with six toes on one foot; a daughter, normal. Third generation: This daughter had five children, amongst whom were a son and daughter each having six fingers on each hand. Fourth generation: The daughter last mentioned had eight children, including one son having six toes on one foot, another son and two daughters each having six fingers on each hand, and one daughter having both six fingers on each hand, and six toes on each foot. Fifth generation: The daughter last mentioned had three children, including a son, doubly deformed like his mother, and a son with six fingers on each hand, the feet being normal. Moreover, one of the two daughters of the fourth generation (with only the hands affected) had eight children, several of whom were normally developed, but the rest were deformed as follows:—One daughter had an osseous thickening at the digital extremity and on the outer border of the fifth metacarpal; one son had six fingers on each hand and six toes on each foot; and another son had six fingers on each hand. Thus, in the

* Quoted from *British Medical Journal*, January 1, 1887.

first generation, one person was affected, and in the second also one; in the third there were two cases of deformity; in the fourth, five; and in the fifth in all five; that is, fourteen deformed persons altogether during five generations of this interesting family."

These specimens may be contrasted with an hypertrophied kidney which received an extra supply of blood owing to the other kidney having been destroyed. This contrast illustrates abnormal proportional growth from inherited conditions, and again from an over-supply of blood.

Many defects in the development of the body are due to the ratio of growth in its parts. The ear is very commonly deformed; such cases often present some brain defect. A boy six years old was brought to me because he did not speak. His movements were controlled by speaking to him, proving that he could hear, but the external ears were symmetrically deformed, the rim at its upper part being absent on each side; further, the uvula was bifid.

In a specimen of a deformed fœtus in the museum, we see many examples of defective or abnormal ratios of bilateral growth. Teratological Catalogue, 321: " The skeleton of a full-timed hydrocephalic fœtus, with extreme shortening of the limbs. All the bones are well ossified except the sternum ; its cartilaginous condition is an indication of arrested development." The bones are abnormally curved. Examples might be multiplied where deformity results from the special quantities of growth in parts. The following may be mentioned :—the underhung jaw, the excessive epicanthic fold, cleft palate, etc.*

In systematic pathological study the seat of histological or functional change enables analogies to be made. The analogies we are going to make are as to the distribution or the parts affected. Many skin diseases are maculated in distribution, the small loci affected are symmetrically or asymmetrically distributed. Among young infants, in the spontaneous movements of many small parts we have evidence that separate scattered loci of brain-tissue are in action, discharging their force to the muscles. Other examples might be given of small loci of tissue undergoing pathological changes. It would be interesting to collect such examples and search for common antecedents. We may place side by side examples of slight movements all

* See paper in *Medical Times and Gazette*, June 21 and 28, and February 11, 1882, on coincident defects of development.

over the body, and cases of freckles, psoriasis, etc., and we may say that both brain-tissue and the skin, tissues allied in their development, inherit a tendency to separate action in small loci of tissue.

It would occupy too much time to follow this subject, the analogy of pathological processes compared as to distribution. The following examples, kinetic and trophic, may be mentioned:—hemiplegia, hemichorea, hemispasm, facial tic and facial hemiatrophy, asymmetry of postures and unequal proportional growth, etc. It is suggested that we should search for the antecedents of the special distribution of a process, as well as for the origin of the kind of action.

The distribution of a disease sometimes enables us to determine its essential antecedent. Atheroma of the arteries is found mainly at such points as bear the greatest mechanical impact and strain.

The part of the body in which some change is occurring may determine the pathological character of the action observed. A change, normal in one part, may, if it occurs elsewhere, be abnormal. The distribution, or seat of action . in a pathological process, is, then, an important subject of study. Some processes are considered pathological when they occur asymmetrically, while when seen equally on both sides of the body they are not considered abnormal. This has already been pointed out in the case of exhaustion of the central nerve-system. Local exhaustion of its tissue is called "hysterical palsy :" if such conditions be generally distributed, and temporary in duration, the state may be called fatigue.

General pigmentation of the skin, say, from sun rays, is not abnormal. If seen in the form of freckles symmetrically distributed it is less liked; if one-sided, it is distinctly a pathological condition.

A widespread loss of fat is less abnormal than loss of fat from one side of the face, or from one limb or from one orbit, as in the case narrated, where it is distinctly pathological.

Time is an attribute of acts of growth which may give them a pathological character, or a character different from the normal. The order of growth and development of the teeth may differ from the average; dentition may be late, early, or it may be that the teeth do not appear in the usual order or series.

With regard to the growth of hair in different parts of the body, it is abnormal if the beard or pubic hair appear before puberty, the growth of beard in the woman of advanced age is perhaps abnormal. I have seen a photograph of an idiot boy ten years old, with full growth of pubic hair; such was certainly abnormal. Grey hair and baldness are abnormal at thirty years of age, not at fifty.

There is a normal relation in the time at which each portion of the cartilaginous skeleton ossifies, and at which it is complete. Premature ossification of the skull is an abnormality seen in microcephalic idiots. In the skulls of some nations ossification is complete at a much earlier age than in the European.

Some processes of growth owe their special characters and attributes to the relation of the time of special stimulation, such processes of growth being abnormal owing to the time of the stimulus. Pressure on growing parts, if constant, tends to atrophy, if intermittent to hypertrophy.*

The process of growth may be unusual or abnormal, because of the relation of its time of action to surrounding forces. Thus buds may appear in cold weather and be nipped by the frost.

In the production of morbid growths, a part of the pathology appears to be the growth of embryonic tissue at a time of life when it is abnormal.

* See Paget and Hunter.

This was pointed out by Mr. Savory in his Bradshaw lecture, 1886.

It was shown that time was one of the principal attributes of movements. Hence the abnormal character of a nerve-muscular act is often a relation as to time : the rapidity of talk, the slowness of response to questions or to impressions, may show mental abnormality. In chorea the time of movements is very different from the normal.

The study of pathological histology has added greatly to our knowledge ; still we have much need of information as to the causes or antecedents of such conditions of growth. It appears, then, to be important whenever we study a pathological growth, to look to the action of those forces which stimulate and control its nutrition. We have now to consider the action of those forces which precede healthy or unhealthy growth respectively in parts of the subject.

Defective stimulation, or the continued non-stimulation, of a certain series of movements in man, tends to loss of that special motor-action, or special mode of co-ordination of nerve-centres. It may be lost from want of use in the individual, and in the course of generations it may be lost in the species. On the contrary, frequent stimulation

K

of an action, whether trophic or kinetic, is apt to strengthen it, and lead to its recurrence, this is supposing that there is a good supply of blood to the part.

Various forces may act as stimuli to nutrition. We can but glance at the subject. *Light* has been shown to have a marked effect in stimulating growth in plants, and in determining the ratios of growth in their parts. Light in man is necessary to prevent anæmia, and many of his kinetic actions are regulated by the sight of objects. In certain crustacea the eyes appear to have been lost in a section of the species which for generations had been living in darkness. An abnormal kinetic condition often follows a visual impression; this may be due to a previous abnormal impressionability of the subject, or to the altogether abnormal character of the stimulus. A servant, going into a room at night, saw a burglar getting in at the window, and fell down in an epileptic fit.

Heat, or as we say, warmth and cold, are important forces in regulating quantities of growth. Warmth aids the healing of wounds, the production of granulations, the nutrition of a paralysed limb, and in some cases aids in lessening inflammation. The effects of heat, as an element of climate affecting growth, are well known both in the case of plants and animals.

The effects of *mechanical stimulus* upon quantities in growth have been illustrated in the case of the Chinese foot kept small by pressure, and in cases of hypertrophy following intermittent pressure.* The white patch near the apex of the heart taken from a rickety chest is a fibrous overgrowth due to constant mechanical irritation. High tension in arteries is followed by atheroma and hypertrophy of the heart.

Some mechanical stimuli appear to be distinctly antecedents of pathological overgrowth. This is seen in the case of "sweep's cancer," or epithelioma of the lip, following the continued irritation of soot. When two or three generations of men follow the same occupation, the cancer appears earlier in the second than in the first generation; the tendency or impressionability appears to be communicated by heredity. This impressionability to trophic changes sequent to special stimulation may be compared with the impressionability of the children of edu‐cated parents to the effects of teaching.

The beneficial effects of rest, or absence of me‐chanical strain, upon some conditions of growth are well known.

* See "Paget's Lectures on Pathology," vol. i. p. 91.

Conditions of pathological growth may apparently be sequents to special abnormal stimuli of a particulate or material kind. We do not know how certain minute particles of matter can produce marked effects upon growth such as we see follow the advent of the protoplasm of the pollen grain to the ovule of the ovary. In the leaf of the oak a great outgrowth of tissue follows the irritation of the egg of the insect producing the common gall ; *Syphilitic virus* appears to convey a stimulus leading to many abnormal forms of overgrowth.

It seems probable that the stimuli producing pathological processes of growth may be forces that acted upon the ancestors of the individual. It seems probable that circumstances in the life-history of the ancestors are very common antecedents of the special ratios of growth. Many examples might be quoted from the works of C. Darwin. I would suggest that, as many examples of proportional growth are undoubtedly thus produced, it is likely that many examples of abnormal ratios of growth are sequential to similar antecedents.

We consider it probable that the tendency to a particular mode of abnormal growth is inherited when similiar abnormalities have occurred in other members of a family; so also when deformities are multiple in the same subject. It has been shown that coincident defects of development in the same subject are common, and that in such cases the brain is often defective likewise.*

In concluding these lectures it may be well to glance over our work, to see how far the objects put forward at the commencement have been fulfilled. We proposed to study movements that we might gain knowledge concerning the action of portions of brain-tissue; from such studies we deduced principles which we proceeded to apply to modes of growth and other vital actions.

Concerning movements, it was shown that their characters depend upon their intrinsic attributes. We were thus enabled in many cases to substitute physical descriptions for metaphysical terms, and we were able to arrange many facts in classes such as seemed adapted to demonstrate their causation. Thus means were found for giving descriptions of the states termed coma, chorea, the neurotic condition in children, rickets, and for defining the signs indicating old age.

A collection of physical signs, indicating condi-

* See paper in *Medical Times*, June, 1882, on " Coincident Defect in Development."

tions of the nerve-system, was mentioned incidentally in these descriptions.

The attempt was made to define a working
hypothesis as to what is meant by "reversion."

Among the principles and modes of procedure
deduced from the study of movements, we found it
necessary to be prepared to make sub-divisions of
the living thing into parts which can act separately;
to note the intrinsic attributes of a vital act, independent of its results or sequents, which often owe
their special characters to surrounding objects. We
also learnt the necessity of trying to determine the
antecedents of every vital act, and in so doing we
obtained much evidence for the maintenance of
the hypothesis that "every vital act, or act of
growth or movement, requires among its antecedents a supply of pabulum, and stimulation by some
force incident to the subject." Further, evidence
was advanced that the supply of pabulum, and the
stimulation of the subject by physical forces could
determine the intrinsic attributes of the action; also
that the special characters were due to its intrinsic
attributes and the surroundings. The special combinations and series of acts often necessarily determined the results, and were themselves controlled
by the forces producing the act.

Pathological and teratological modes of growth were shown to be largely due to their intrinsic attributes, as was seen in the study of proportional growth, bilateral growth, and their consequences. It appeared that the stimulus producing an act, might itself be so unusual in kind as to produce an abnormal result.

In studying the overgrowth or undergrowth of parts from overuse or disuse, it appeared that this was really due, not to the voluntary character of the action, but to the quantity of stimulation incident to the part affected.

Incomplete as this treatise is, still I trust that some good has been effected, in pointing to lines of observation and thought which have guided much of my own work.

PRINTED BY WILLIAM CLOWES AND SONS, LIMITED, LONDON AND BECCLES.